PR[RAINBOW AND THE PHILADELPHIA EXPERIMENT

David Hatcher Childress

Adventures Unlimited Press

PROJECT RAINBOW AND THE PHILADELPHIA EXPERIMENT

Morris K. Jessup and US Navy Teleportation

Project Rainbow and the Philadelphia Experiment

Copyright © 2024
by David Hatcher Childress

ISBN 978-1-948803-71-7

Published by:
Adventures Unlimited Press
One Adventure Place
Kempton, Illinois 60946 USA
auphq@frontiernet.net

Cover by Terry Lamb

AdventuresUnlimitedPress.com

10 9 8 7 6 5 4 3 2 1

Dedicated to the scientists and philosophers whose quest for knowledge takes them to the most distant stars and to the innermost worlds.

When a distinguished but elderly scientist states that something is possible, he is almost certainly right. When he states that something is impossible, he is probably wrong.
—Arthur C. Clarke

Actually the biggest deterrent to scientific progress is a refusal of some people, including scientists, to beleive that things that seem amazing can really happen.
—George S. Trimble, Former NASA director

My own special interest in the Philadelphia Experiment was connected with the possibility that a shift in the molecular composition of matter, induced by intensified and resonant magnetism, could cause an object to vanish—one possible explanation of some of the disappearances within the Bermuda Triangle.
—Charles Berlitz,
The Philadelphia Experiment, p. 14.

PROJECT RAINBOW AND THE PHILADELPHIA EXPERIMENT

Morris K. Jessup and US Navy Teleportation

A photo of the USS *Eldrige* at sea in 1944, during WWII.

TABLE OF CONTENTS

One of the few photos of Morris K. Jessup.

Chapter 1
The Philadelphia Experiment

There has to be an invisible sun
That gives us hope when the whole day's done
It's dark all day and it glows all night
—*Invisible Sun*, The Police

Let us set the scene. It is the early 1950s and flying saucers are in the news. Hollywood is producing movies and serials like *Radar Men from the Moon, Disc Men from Mars, Earth vs the Flying Saucers* and such. Buck Rogers and Flash Gordon were zipping around in their buzzing electric spaceships and alien worlds were the subject of many comic books.

Into this world came an author and researcher named Morris K. Jessup. He was born in Iowa and went to school in Michigan. He went to Peru and Brazil in the 1920s and 30s. He became fascinated by the stonework in Cuzco, Peru, and other ancient cities around Cuzco. After WWII he became interested in UFOs and wrote several books about flying saucers.

Then he became entangled in Project Rainbow or what is called the Philadelphia Experiment. He was exposed to fantastic science and strange ideas. Then he was found dead in Florida under suspicious circumstances. His death was as mysterious as his life.

So, what was the Philadelphia Experiment and who was Morris Jessup?

The experiment was allegedly based on an aspect of unified field theory, a term coined by Albert Einstein to describe a class of potential theories; such theories would aim to describe—mathematically and physically—the interrelated nature of the forces of electromagnetism and gravity, in other words, uniting their respective fields into a single field.

In a rather technical article Wikipedia says this about the

unified field theory:

The first successful classical unified field theory was developed by James Clerk Maxwell. In 1820, Hans Christian Ørsted discovered that electric currents exerted forces on magnets, while in 1831, Michael Faraday made the observation that time-varying magnetic fields could induce electric currents. Until then, electricity and magnetism had been thought of as unrelated phenomena. In 1864, Maxwell published his famous paper on a dynamical theory of the electromagnetic field. This was the first example of a theory that was able to encompass previously separate field theories (namely electricity and magnetism) to provide a unifying theory of electromagnetism. By 1905, Albert Einstein had used the constancy of the speed of light in Maxwell's theory to unify our notions of space and time into an entity we now call spacetime and in 1915 he expanded this theory of special relativity to a description of gravity, general relativity, using a field to describe the curving geometry of four-dimensional spacetime.

In the years following the creation of the general theory, a large number of physicists and mathematicians enthusiastically participated in the attempt to unify the then-known fundamental interactions. Given later developments in this domain, of particular interest are the theories of Hermann Weyl of 1919, who introduced the concept of an (electromagnetic) gauge field in a classical field theory and, two years later, that of Theodor Kaluza, who extended General Relativity to five dimensions. Continuing in this latter direction, Oscar Klein proposed in 1926 that the fourth spatial dimension be curled up into a small, unobserved circle. In Kaluza–Klein theory, the gravitational curvature of the extra spatial direction behaves as an additional force similar to electromagnetism. These and other models of electromagnetism and gravity were pursued by Albert Einstein in his attempts at a classical unified field theory. By 1930 Einstein had already considered the Einstein-Maxwell–Dirac System

12

[Dongen]. This system is (heuristically) the super-classical [Varadarajan] limit of (the not mathematically well-defined) quantum electrodynamics. One can extend this system to include the weak and strong nuclear forces to get the Einstein–Yang-Mills–Dirac System. The French physicist Marie-Antoinette Tonnelat published a paper in the early 1940s on the standard commutation relations for the quantized spin-2 field. She continued this work in collaboration with Erwin Schrödinger after World War II. In the 1960s Mendel Sachs proposed a generally covariant field theory that did not require recourse to renormalization or perturbation theory. In 1965, Tonnelat published a book on the state of research on unified field theories.

The article then concludes:

Theoretical physicists have not yet formulated a widely accepted, consistent theory that combines general relativity and quantum mechanics to form a theory of everything. Trying to combine the graviton with the strong and electroweak interactions leads to fundamental difficulties and the resulting theory is not renormalizable. The incompatibility of the two theories remains an outstanding problem in the field of physics.

So, the unified field theory of gravity, magnetism and electricity has come to a major stall because of differences inthe theories of general relativity and quantum mechanics. One has to do with "fields" of energy and the other "particles" of energy. To simplify this concept we can use the example of sunlight hitting the earth creating daylight. Is the sunlight energy from the sun hitting the planet a wave of electromagnetic energy or as particles hitting us from space? The answer here seems to be the former: that energy from the sun hits us as a wave of energy, not as particles.

According to some accounts, unspecified "researchers" thought that some version of this field would enable using large electrical generators to bend light around an object via refraction, so that the object became completely invisible. The Navy regarded this as being of military value and it sponsored the Philadelphia Experiment.

Another unattributed version of the story proposes that researchers were preparing magnetic and gravitational measurements of the seafloor to detect anomalies, supposedly based on Einstein's attempts to understand gravity. In this version, there were also related secret experiments in Nazi Germany to find anti-gravity, allegedly led by SS-Obergruppenführer Hans Kammler.

There are no reliable, attributable accounts, but in most accounts of the supposed experiment, USS *Eldridge* was fitted with the required equipment at the Philadelphia Naval Shipyard. Testing began in the summer of 1943, and it was supposedly successful to a limited extent. One test resulted in *Eldridge* being rendered nearly invisible with some witnesses reporting a "greenish fog" appearing in its place. Crew members complained of severe nausea afterwards.

Also, reportedly, when the ship reappeared, some sailors were embedded in the metal structures of the ship, including one sailor who ended up on a deck level below where he began and had his hand embedded in the steel hull of the ship as well as some sailors who went "completely bananas." There is also a claim the experiment was altered after that point at the request of the Navy, limiting it to creating a stealth technology that would render USS *Eldridge* invisible to radar. None of these allegations have been independently substantiated.

Other versions of the story give the date of the experiment as October 28, 1943. In this version, *Eldridge* not only became invisible, but disappeared from the area and teleported to Norfolk, Virginia, over 200 miles (320 km) away. It is claimed that the *Eldridge* sat for some time in view of men aboard the ship SS *Andrew Furuseth*, whereupon *Eldridge* vanished and then reappeared in Philadelphia at the site it had originally occupied.

Many versions of the tale include descriptions of serious side effects for the crew. Some crew members were said to have been physically fused to bulkheads while others suffered from mental disorders, some re-materialized inside out, and still others vanished. It is also claimed that the ship's crew may have been subjected to

14

brainwashing to maintain the secrecy of the experiment.

So, what is the real story? The US Navy itself apparently became very interested in the story of Project Rainbow and the Philadelphia Experiment. Why was that? Was Jessup actually murdered? And if this was the case, by whom and for what reason? Let us now look at Carl or Carlos Allende. We will take a close look at Jessup later in the book.

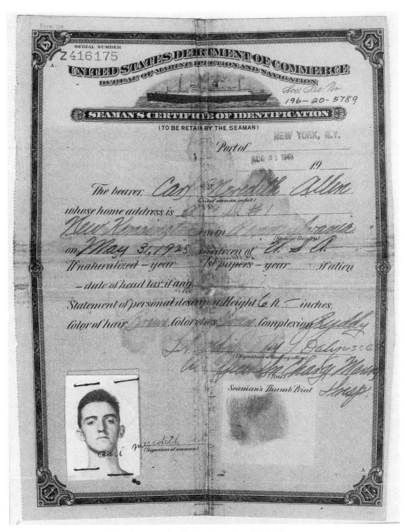

The Seaman's Certificate for Carl Allen.

Chapter 2

The Mysterious Carl M. Allen

Oh tell me where your freedom lies
The streets are fields that never die
Deliver me from reasons why
You'd rather cry, I'd rather fly
—*The Crystal Ship,* The Doors

In many ways the central figure in the Project Rainbow/ Philadelphia Experiment was a rather mysterious person named Carl Allen who often called himself Carlos Allende.

Carl Meredith Allen was born on May 31, 1925 in Springdale, Pennsylvania. He died on March 5, 1994, at age 68 in Greeley, Colorado.

Apparently, Allen's family described him as brilliant in school with a "fantastic mind" but also as a person who never held any particular job for long and was a drifter. He was also known as a "master leg-puller," pulling pranks on people, or making up stories to get out of work in general. In 1942 he joined the US Marine Corps but was discharged within that year.

Right after that he enlisted in the United States Merchant Marine, at first serving on the SS *Andrew Furuseth* and then many other ships until 1952 when he left service. During World War II, nearly 250,000 civilian merchant mariners served as part of the US military, transporting supplies and personnel.

Allen would later claim that in 1943 he witnessed an invisibility experiment carried out by the US Navy at the Philadelphia Naval Shipyard. He also claimed that he met Albert Einstein there, and for several weeks, was schooled in physics by Einstein. At that time Einstein was a professor at nearby Princeton University. Allen was definitely schooled in theoretical physics, particularly the Unified Field Theory, as is evident in his various letters and other writings.

During his lifetime Allen would use many aliases including Carlos Miguel Allende, Senor Professor and Colonel Carlos Miguel Christofero Allende. One time when he wrote to the rocket engineer Wernher von Braun he called himself Dr. Karl Merditt Allenstein. Allen turned up in various places including New Jersey, Arizona, Colorado, and Mexico. He eventually ended up in Greeley, Colorado where he died at the Centennial Health Care Center. His obituary was published in the *Greeley Tribune* on March 8, 1994. The obituary said that he was born in Springdale, Pennsylvania and had come to Greeley in 1986. It made no mention of Project Rainbow, the Philadelphia Experiment or Allen's fame and notoriety in these areas. It is not known how Allen earned a living or what jobs he might have done. He may have lived on welfare and social security payments.

The Annotated Edition of *The Case for the UFO*

In late 1955 an anonymous package arrived at the US Office of Naval Research (ONR). It contained a copy of Morris K. Jessup's book *The Case for the UFO: Unidentified Flying Objects* that was filled with handwritten notes in its margins, written with three different shades of blue ink, appearing to detail a debate among three individuals. They discussed ideas about the propulsion for flying saucers, alien races, and expressed concern that Jessup was

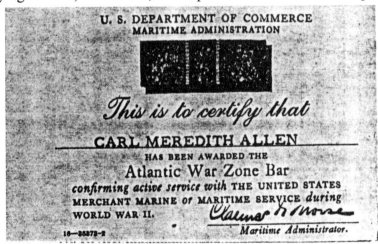

Carl Allen's Navy ID.

too close to discovering their technology.

When Jessup was invited to the Office of Naval Research a year later and shown the annotated copy of his book, he noticed the handwriting of the annotations resembled a series of letters he received from Carl Allen, who also signed some of his letters "Carlos Miguel Allende." In the letters to Jessup, Allen put forward a story of dangerous science based on unpublished theories by Albert Einstein that had been put into practice at the Philadelphia Naval Shipyard in October 1943.

The ONR was quite interested in this annotated edition of Jessup's book. Apparently they were intrigued by the notes in the book and the authoritative manner in which the three "entities" spoke in their conversations about the text. It is this commentary from Allen that provoked a profound response from the Office of Naval Intelligence; it prompted them to create their own special edition of Jessup's *The Case for the UFO* using a private printing press in Texas. We will discuss this at length in the next chapter.

Allen's scribbled commentaries in Jessup's book suddenly took on a life of their own when the Navy asked the Varo Manufacturing Corporation of Garland, Texas, who did contract work for ONR, to produce mimeographed copies of the book with Allen's annotations in color, to preserve the different colored pens used by the three "commentators."

In 1956 the ONR attempted to locate "Carlos Allende" but failed. It is thought that Carl Allen was in Mexico or Minnesota at the time, but no one really knows his travels in the late 50s and early 60s.

Sometime in 1956 or so, a dozen and eventually 127 copies of the "Varo Edition" were printed. This annotated edition—from the US Navy—became the heart of many "Philadelphia Experiment" books, documentaries, and movies to come.

Carl Allen's Letters to Morris Jessup

As previously mentioned, when Jessup was invited to the Office of Naval Research in 1957 and shown the annotated copy of his book, he noticed the handwriting of the annotations resembled a series of letters he had received from Carl Allen, who also signed some of his letters "Carlos Miguel Allende." As mentioned

19

earlier, Allen put forward a story of dangerous science based on the Unified Field by Albert Einstein that had been put into practice at the Philadelphia Naval Shipyard in October 1943.

As Allen had sent an annotated edition of Jessup's book to the Navy, he also began sending letters at the same time to Morris K. Jessup. His first letter, which concerned levitation, is apparently lost. But Jessup received a second letter, now called the "first letter," mailed to Jessup sometime in 1956. It reads:

Carlos Miguel Allende
R. No. 1, Box 223
New Kensington, Pennsylvania

My Dear Dr. Jessup,

Your invocation to the public that they move, en masse, upon their representatives and place enough pressure at the right and sufficient number of places, such that a new law demanding research into Dr. Albert Einstein's Unified Field Theory may be enacted, is not at all necessary. It may interest you to know that the good doctor was not so much influenced in his retraction of that work by mathematics, but by "humantics" (humanism).

Einstein's later computations, done strictly for his own edification and amusement, and based on cycles of human civilization and progress compared to the growth of man's general overall character, were enough to horrify him. Thus, we are "told" today that the Unified Field Theory was " incomplete."

Dr. Bertrand Russell asserts privately that it *is* complete. He also says that mankind is not ready for it, and shan't be until after WWII. Nevertheless, the "results" of my friend, Dr. Franklin Reno, were used. These were a complete recheck of that theory, with a view to any and every possible quick use of it, if feasible in a very short time. There were good results—a group math re-check and good physical "result" to boot. Yet, the Navy fears using this result. The result stands today as proof that the Unified

Field Theory, to a certain extent, is correct. Beyond that, no person in his right senses, or having any senses at all, will dare go.

I am sorry that I misled you in my previous missive. True enough, such a form of levitation has been accomplished as described. It is also a very commonly observed reaction of certain metals to certain fields surrounding a current. Had Faraday concerned himself with the magnetic field surrounding an electric current, we would not have this current time-bombish, "ticking off towards destruction" geopolitical atmosphere.

Alright, alight! The "result" was complete invisibility of a ship. Destroyer type, and all of its crew, while at sea (October 1943). The "field" used was effective in an oblate spheroidal shape, extending one hundred yards out from each end of the ship. Any person within that sphere became vague in form, and observed others onboard in the same state—walking upon nothing. Any person outside that sphere could see nothing, save the clearly defined shape of the ship's hull in the water—provided that person was just close enough to see, yet barely outside of, that field.

Why tell you this now? It's very simple. If you choose to go mad, then you will reveal this information. Half of the officers and crew of that ship are, at present, mad as hatters. A few are even confined to certain areas where they may receive trained, scientific aid when they either "go blank" or "go blank and get stuck."

"Going blank" is not at all an unpleasant experience to healthily curious sailors. It is when they also "get stuck" that it becomes "Hell Incorporated." The man thusly stricken cannot move of his own volition, unless two or more of those who are within the field go and touch him, quickly—else he "freezes."

If a man freezes, his position must be marked out carefully, and then the "field" is cut-off. Everyone but that "frozen" man is able to move—to appreciate apparent solidity again. Then, the newest member of the crew must find a spot on the frozen man's face or bare skin that is not covered by usual uniform clothing.

Sometimes, it takes only an hour or so; sometimes it takes all night and all day, or worse. It once took 6 months to get a man "unfrozen." This "deep freeze" is not psychological. It is the result of a hyper-field that is set up within the field of the body when the "scorch" field is turned on.

A highly complicated piece of equipment had to be constructed in order to unfreeze those who became "true froze" or "deep froze" subjects. Usually a "deep freeze" man goes stark-raving, gibbering, running mad if his "freeze" is for more than a day.

I speak of time, for deeply "frozen men" are not aware of time as we know it. They are like semi-comatose persons who live, breathe, look, and feel, but still are unaware—in a "netherworld." A man in an ordinary, common freeze is aware of time—sometimes cutely so. Yet they are never aware of time, as you or I are aware of it. The first "deep freeze" took 6 months to rectify. It also took over 5 million dollars worth of electronic equipment and a special ship berth.

If, around or near the Philadelphia Navy Yard, you see a group of sailors in the act of putting their hands upon a fellow, or upon "thin air," observe the digits and appendages of the stricken man. If they seem to waver as though within a heat-mirage, go quickly and put your hands upon him, for that man is the most desperate of men. Not one of those men ever wants to become invisible again. I do not think that much more need be said as to why mankind is not ready for forcefield work.

You will hear phrases from these men such as "caught in the flow (or the push)," "stuck in the green," "stuck in molasses," or "I was going fast." These refer to some of the after-effects of forcefield work. "Caught in the flow" describes exactly the "stuck in molasses" sensation of a man going into a "deep freeze" or "plain freeze." "Caught in the push" can either refer to that which a man feels briefly when he is either about to inadvertently "go blank" (i.e., become invisible) or about to "get stuck" in a "deep freeze" or "plain freeze."

There are only a very few of the original crew left now. Most went insane. One just walked "through" the wall, in sight of his wife, child, and two other crewmembers, and was never seen again. Two "went into the flame" (i.e., "froze" and caught fire) while carrying common boat compasses. One man carried the compass and caught fire. The other came for the "laying on of hands," as he was nearest; but he, too, took fire. *They burned for 18 days.* The faith in "hand laying" died when this happened, and men's minds went by the score. The *experiment* was a complete success, but the *men* became complete failures.

Check the 1944 Philadelphia papers for a tiny, one-paragraph item describing the sailors' actions after their initial voyage. They "raided" a Navy Yard beer joint and caused such shock and paralysis in the waitresses that little comprehensible could be gotten from them, save that paragraph, The writer does not believe it, and says, "I only wrote what I heard; them dames is daffy. So, all I got was a bedtime story."

Check the observer ship's crew—a Matson Lines Liberty ship out of Norfolk, the SS *Andrew Furuseth*. The Chief Mate was named Maundsley. (The company may have the ship's log for that voyage, or the Coast Guard may have it.) One crewmember, Richard "Splicey" Price, may remember other names of deck crewmen. (The Coast Guard has records of sailors issued "papers.") Mr. Price was 18 or 19 in October 1943, and lived at his old family home in Roanoke, Virginia—a small town with a small phone book. These men were witnesses. I ask you to do this bit of research, simply that you may choke on your tongue when you remember what you have "appealed by made law."

Very Disrespectfully Yours,

Carl M. Allen

P.S. Will help more, if you see where I can.

The above letter is known as the first letter to Jessup. Days later, Jessup received an additional letter which said simply:

Notes in addition to, and pertaining to, prior missive.
(Contact Rear Admiral Rawson Bennett, Navy Chief of Research, for verification of information herein. He may even offer you a job.)
Coldly and analytically speaking, without the howling, I will say the following in all fairness to you and to science: 1) the Navy did not know that the men could become invisible even when not upon the ship or under the field's direct influence; 2) the Navy did not know that men would die from odd effects within or upon the hyper-field; 3) even now, they do not know why this happened. Myself, I "feel" that something pertaining to that common boat compass "triggered" off "the flames." I have no proof, but neither does the Navy.

The experimental ship disappeared from its Philadelphia dock. Only a few minutes later, it appeared at its other dock in the Norfolk area. But the ship again disappeared, and went back to its Philadelphia dock in only a few minutes. This was noted in the newspapers.

To the Navy, this whole thing was so impractical, due to its morale-busting effects, which hindered the efficient operation of the ship. After this occurrence, it was shown that even the basic operation of a ship could not be counted upon. In short, ignorance of this thing bred such terrors that it was deemed too impossible, impracticable, and horrible.

I believe that had you then been working on the project, "the flames" would not have been so unexpected, or such a terrifying "mystery." More than likely, none of these other occurrences would have happened. They could have been prevented by a far more cautious selection of personnel.

Such was not the case. The Navy used whatever "human material" was at hand, without any thought as to character and personality. (If more care were taken in the selection of officers and crew, and if those hob-nailed, Navy-issued shoes were replaced, I would feel that some progress had been accomplished.)

The records of the U.S. Maritime Service House, Norfolk, Virginia (for graduated seamen of their schools), will reveal just who was assigned to SS *Andrew Furuseth* for the months of September and October 1943. I remember positively of one other observers who stood beside me when the tests were going on. He was from New England, and had blond, curly hair and blue eyes. I don't remember his name. I leave it up to you to decide if further work shall be put into this or not, but write in hope that there will be.

Very Sincerely,

Carl M. Allen

Jessup received a third letter from Carl Allen, postmarked DuBois, Pennsylvania, on May 25, 1956:

Having just recently gotten home from my long travels around the country, I find that you have dropped me a card! You ask that I write you "at once." After taking everything into consideration, I have decided to do so. You ask me for what is tantamount to "positive proof" of something that only the duplication of those devices that produced "this phenomenon " could give you...

Mr. Jessup, I could never possibly satisfy such an attitude, nor would the Navy Research Department (then under the present boss of the Navy, Burke) ever let it be known that any such thing was ever done. It was because of Burke's curiosity, willingness, and prompting that this experiment was carried out in the first place. It proved to be a white elephant, but his attitude towards advanced and ultra-advanced types of research is just the thing that put him where he is today. Were the stench of such an experiment ever to come out, he would be crucified.

However, I have noticed that through the ages, once the vulgar passions have cooled-off, these crucified ones achieve something akin to sainthood. You say that this is all "of the greatest importance." I disagree with you. My disagreement is based upon philosophical morality. I could be of some positive help to you, but to do so would require a hypnotist, sodium pentathol, a tape recorder, and an excellent typist-secretary.

As you know, one who is both hypnotized and given "truth serum" cannot lie. I could thus be enabled to not only recall complete names, but also addresses, telephone numbers, and perhaps the very important Z numbers of those sailors I came into contact with. You are bound to get good results. The mind does not *ever* forget. It is my belief that if the Air Force were confronted with such evidence, there would be an uproar. And then there would be a quiet, determined effort to safely achieve that which the Navy failed to achieve.

They did not fail to achieve metallic and organic invisibility, however. I believe that further experiments would produce instant, controlled transport of great tonnages to desired points. Newspaper morgues will divulge even more positive proof of this experiment.

The name of the reporter who skeptically wrote of these incidents (of the barroom brawl with invisible crewmen) can be found, and his and the waitresses' testimony can be added to the records. Once on this track, I believe that you can uncover considerably more evidence to out this scandal.

I do hope you will consider this plan. The ultimate end will be a truth too huge and fantastic not to be told, but it will be backed up by proof positive.

Find where these sailors live now. These folks have a very high "psi" factor in their make-up, which can be intensified under stress, strain, or extreme fright. It also can be re-intensified by hypnosis, thus making it as easy to "read" them as is it to read *The Encyclopedia Britannica*. Even though that barroom raid was staged by partly invisible men, those men can still "see" each other. A check of naval hospital or prison records may reveal the names of these men. Wouldn't you like to speak to some of these men?

Maybe I am suggesting something too methodical for your taste, but I want to crack this thing wide open. My reasons are simply to enable more work to be done upon the Unified Field Theory.

I am a stargazer, Mr. Jessup. I make no bones about this. I feel that if handled properly, man can go where he only now dreams of going. Perhaps the Navy has already used this "accident of transport" to build your UFOs. It is a logical advance from any standpoint. What do you think?

Very Respectfully,

Carl M. Allen

These are the three letters that we have that Allen/Allende wrote to Morris Jessup. They seem very sincere and knowledgeable about Project Rainbow and the Philadelphia Experiment. How much of it is real and how much of it is fantasy, and perhaps mind control manipulation, is speculative.

But there are a number of interesting bits of information to look at here, and it must be said that Carl Allen seemed to believe much of what he was writing about and genuinely believed that there had been some sort of time travel or invisibility experiment involving the USS *Eldridge* that he had witnessed.

The Missing Newspaper Article

Allen claimed to have witnessed this experiment while serving aboard the SS *Andrew Furuseth*. In Allen's account, a destroyer escort was successfully made invisible, but the ship inexplicably teleported to Norfolk, Virginia for several minutes, and then reappeared in the Philadelphia yard. The ship's crew was supposed to have suffered various side effects, including insanity, intangibility, and being "frozen" in place.

When Jessup wrote back requesting more information to corroborate his story Allen said his memory would have to be recovered and referred Jessup to a Philadelphia newspaper article that Allen claimed covered the incident. Allen/Allende wrote:

Check the 1944 Philadelphia papers for a tiny, one-paragraph item describing the sailors' actions after their initial voyage. They "raided" a Navy Yard beer joint and caused such shock and paralysis in the waitresses that little comprehensible could be gotten from them, save that paragraph, The writer does not believe it, and says, "I only wrote what I heard; them dames is daffy. So, all I got was a bedtime story."

While it is claimed that this was a nonexistent article that Allen simply fabricated, evidence of such an article has turned up. According William Moore in his book *The Philadelphia Experiment*[7] he was sent a photocopy of a newspaper article that appears to corroborate the one that Allen had read and perhaps had in his possession. Moore says that the article appears to be referring to an incident in Philadelphia during the war but does not have a date or the name of the newspaper on it. Moore says that the photocopy rests in a "secure safety deposit box" and was sent to him by an anonymous source. The curious article, apparently from 1943, reads:

Strange Circumstances Surround Tavern Brawl

Several city police officers responding to a call to aid members of the Navy Shore Patrol in breaking up a tavern brawl near the US Navy docks here last night got something a surprise when they arrived on the scene to find the place empty of customers. According to a pair of very nervous waitresses, the Shore Patrol had arrived first and cleared the place out—but not before two of the sailors involved allegedly did a disappearing act. "They just sort of vanished into thin air... right there," reported one of the frightened hostesses, "and I ain't been drinking either!" At that point, according to her account, the Shore Patrol proceeded to hustle everybody out of the place in short order.

A subsequent chat with the local police precinct left no doubts as to the fact that some sort of general brawl

Strange Circumstances Surround Tavern Brawl

Several city police officers responding to a call to aid members of the Navy Shore Patrol in breaking up a tavern brawl near the U.S. Navy docks here last night got something of a surprise when they arrived on the scene to find the place empty of customers. According to a pair of very nervous waitresses, the Shore Patrol had arrived first and cleared the place out —but not before two of the sailors involved allegedly did a disappearing act. "They just sort of vanished into thin air . . . right there," reported one of the frightened hostesses, "and I ain't been drinking either!" At that point, according to her account, the Shore Patrol proceeded to hustle everybody out of the place in short order.

A subsequent chat with the local police precinct left no doubts as to the fact that some sort of general brawl had indeed occurred in the vicinity of the dockyards at about eleven o'clock last night, but neither confirmation nor denial of the stranger aspects of the story could be immediately obtained. One reported witness succinctly summed up the affair by dismissing it as nothing more than "a lot of hooey from them daffy dames down there," who, he went on to say, were probably just looking for some free publicity.

Damage to the tavern was estimated to be in the vicinity of six hundred dollars.

The newspaper story referred to by Carl Allen.

had indeed occurred in the vicinity of the dockyards at about eleven o'clock last night, but neither confirmation nor denial of the stranger aspects of the story could be immediately obtained. One reported witness succinctly summed up the affair by dismissing it a nothing more than "a lot of hooey from them daffy dames down there," who, he went on to say, were probably just looking for some free publicity.

Damage to the tavern was estimated to be in the vicinity of six hundred dollars.

So, we have a curious newspaper article that appears to be genuine. It does not actually validate the Project Rainbow Philadelphia Experiment but it is a curious story about invisibility and the US Navy. Was this brawl caused by sailors being suddenly transported to the tavern or nearby and then suddenly yanked from the tavern as the experiment shifted from one location to another?

The Hefferlin Manuscript

Moore also has a curious mention near the end of his book about "The Hefferlin Manuscript" and suggests that this might have been something that Carl Allen had read. Moore says in looking for newspaper and magazine articles from the 1940s that may have referenced the Project Rainbow and the Philadelphia Experiment he found a couple of interesting items:

One such source is an obscure and rather bizarre tract entitled "The Hefferlin Manuscript" which apparently antedates the Allende letters. Mentioned therein is an illustrated *Life* article supposedly from sometime during the 1940s which allegedly dealt with experiments in invisibility conducted by Hungarian scientists during the early days of World War II. Although the account is unclear about whether these took place in Hungary or the United States, it is interesting to note that Dr. John von Neumann... was born in Budapest in 1903.

"The Hefferlin Manuscript" is a 30-page typed and

mimeographed manuscript that was published by a California esoteric book and newsletter publisher called Borderland Sciences. It was apparently published around 1960 and is a collection of missives compiled Gladys Hefferlin who lived in Livingston, Montana as well as San Francisco. These missives originated in 1947 from Mr. W.C. Hefferlin and Gladys Hefferlin, who channeled much of the information through "Controlled Mental Communication." Mr. Hefferlin is also said at the beginning to have designed a circular aircraft in the late 1930s.

The manuscript starts with the couple meeting a man in San Francisco who teaches them telepathy. It then goes on to say circular aircraft (flying saucers) were being made at a secret place in Tibet which they call Shangri La. It says that 350 of the craft have been built and they are flying all over the world.

The manuscript then describes the "Ancient Three" who supposedly were active in different ways during World War II manipulating some of the activity behind the scenes. The manuscript goes on to describe "Rainbow City" which is an ancient city—millions of years old—that is in Antarctica and is made of plastic. The city is inhabited by humans that are seven-and-a-half-feet-tall. There are some serpent people there as well.

It then goes on about "Portals" for a number of pages. Says the manuscript on page 12:

> "Portals" are a means of entering or traveling between places thousands of miles distant; these are used also for tracing back through the past, all done by space warpage. These Portals are mentally or manually controlled.

These portals can go to any place in the world and a person can travel instantly from place to place through these portals. One can even use these portals to snatch an object from a place, like a hand reaching through a porthole, or deliver a foreign object instantly, like a coin or a small statuette. Says the manuscript on page 24:

> We used the Portals to transport material and men from and to many places on earth. These same Portals will reach out through local space to the Moon, but not much farther

at present.

…The Portals have been in almost steady use for most of the time our group has been in the Antarctic. A great deal of shuttle service was to our base in the Himalayas; we call it Shangri La Valley. When Hungary was taken over by the Germans the then Regent of Hungary and his family were moved to Shangri La and then to the Antarctic Rainbow City by one of the Portals.

Early in World War II the scientists of Hungary who had developed the Invisibility Ray as pictured in *Life* Magazine, also a Paralysis Ray, joined our group by plane to Shangri La. Later, they, with their families, were moved by Portal to Rainbow City.

…The rumored Japanese tunnel to Korea was started by captured slave-labor; but it was checked and blown up by use of the Portals a form of atom bomb developed by the scientists.

The manuscript then talks about a train system, which is different from the Portals, out of the Rainbow City in Antarctica that goes to certain terminals around the world. One tunnel goes to a swamp in the middle of South America where the ruins of a great seaport can be seen. It mentions explorers going into the region and never returning. Another tunnel goes to an entrance on an Indian Reservation in the American Southwest, probably on the Navajo Nation. Another tunnel entrance is in western Wyoming. Other tunnels go to all the continents. The tunnels are lighted by a "cold light system." Some tunnels go under the ocean to ruins of Atlantis and Lemuria, which they call Le Muria. Machinery is only found in the terminals underground and not at the entrances to the tunnels.

On page 28 of the manuscript it is claimed that the lost treasure of the Incas is within one of these tunnels, but the people of Rainbow City have left it in place.

In the last two pages of the manuscript (pages 29-30) an interesting claim is made about a Portal entrance on the ocean floor midway between San Francisco and Hawaii. This portal had a machine "used to transport these material substances between

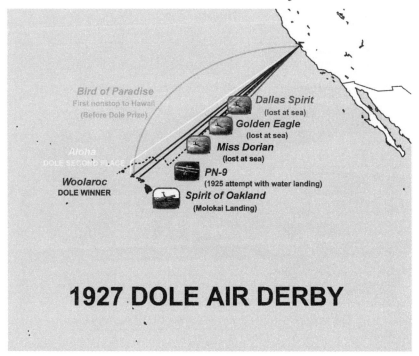

Bird of Paradise
First nonstop to Hawaii
(Before Dole Prize)

Dallas Spirit
(lost at sea)

Golden Eagle
(lost at sea)

Miss Dorian
(lost at sea)

Aloha
DOLE SECOND PLACE

Woolaroc
DOLE WINNER

PN-9
(1925 attempt with water landing)

Spirit of Oakland
(Molokai Landing)

1927 DOLE AIR DERBY

the earth and Moon. Living organic matter was carried by Space Ships traversing the space between earth and Moon."

The manuscript ends with an interesting supposition that this Portal in the Pacific Ocean is still active and caused the disappearance of three planes during what it calls the "Dole Flight." This is very similar to the suggestion of an interdimensional portal in the "Bermuda Triangle," which is in the Atlantic.

The "Dole Flight" mentioned is a real event and it is known as the Dole Air Race or the Dole Derby. It was an air race across the Pacific Ocean from Oakland, California to Honolulu held in August 1927 that resulted in several deaths and vanished airplanes. It was sponsored by the Dole Pineapple Company.

There were eighteen official and unofficial entrants; of them fifteen officially drew for starting positions, and of those fifteen, three aircraft crashed resulting in three deaths before the race. Eight aircraft eventually started the race on August 16, 1927. Only two successfully arrivied in Hawaii; *Woolaroc*, piloted by Arthur Goebel and William Davis, arrived after a 26 hour, 15 minute flight, leading runner-up *Aloha* by two hours.

Of the other six aircraft, two had crashed on takeoff, two

The airplane named *Golden Eagle* the 1927 Dole Air Derby.

were forced to return for repairs, and two went missing during the race (*Golden Eagle* and *Miss Doran*). One of the aircraft that was repaired, the *Dallas Spirit* took off again to search for the missing aircraft several days later and also vanished over the sea. In all, before, during, and after the race, ten people died and three airplanes vanished during the race.

Neither *Golden Eagle*, *Miss Doran* or *Dallas Spirit* was ever seen again. Dole put up a US$10,000 reward for anyone who found any of the planes; this was matched by each plane's sponsors, for a total of US$20,000 reward for each aircraft.

Golden Eagle was last reported approximately halfway to Honolulu, at the southern edge of the course with Aloha at 2 am PST on August 17 by SS *City of Los Angeles*, relaying messages from the Army Signal Corps. Because *Golden Eagle* had sufficient fuel to reach and pass Hawaii altogether, some theorized the aircraft may have overshot the goal in the darkness; but most rejected that theory as ludicrous, as the *Golden Eagle* would have passed over the islands in broad daylight.

A total of 42 Navy ships were involved in the search for the *Golden Eagle* and *Miss Doran*, joined by smaller vessels based in

Hawaii. USS *Langley* and *Aroostook* departed from San Diego the night of August 17. The search fleet also included three submarines.

After repairs *Dallas Spirit* left Oakland for Honolulu on August 19 at approximately 2:20 pm PST. Their last radio message, received at 9 pm that night, was that they were in a tailspin approximately 600 miles (970 km) out to sea. *Dallas Spirit* was also never seen again. Erwin's aircraft was reportedly seen 600 miles (970 km) from Hawaii at noon on August 20, but the report was not confirmed. USS *Hazelwood*, which was approximately 100 miles (160 km) away and searching for the other two missing aircraft when *Dallas Spirit* went down, proceeded to the spot where the last transmission was made, arriving less than three hours later. Hazelwood scoured an area of 3,200 square miles (8,300 km²) without finding any wreckage or flotsam from *Dallas Spirit*. By August 22, naval ships from California had met their counterparts who had left from Hawaii, with neither fleet finding a trace of the three planes.

What the authors of "The Hefferlin Manuscript" believed in 1947 was that the disappearance of these three planes—Bermuda Triangle style—was the result of the Le Muria-Moon Portal that presumably sent these aircraft into outer space, rather than dragging them to the bottom of the sea.

The airplane named *Dallas Spirit* in the 1927 Dole Air Derby.

The manuscript is sort of an outline for a number of stories for the popular magazine *Amazing Stories* which at the time featured the Shaver Mystery stories that included time machines, Lemuria, portals, underground worlds, thought rays, and incredible technology. Indeed, at the beginning of "The Hefflerin Manuscript" there is a discussion of exactly this: the Shaver Mystery and the many issues of *Amazing Stories* that told these stories as facts.

"The Hefferlin Manuscript" is science fiction of the "channeled" sort and is a mishmash of various *Amazing Stories* topics and genuine history such as its strange discussions about WWII and the Dole Air Race. But it is remarkable in the early time frame—1947—from which it came and its discussion of topics that would soon be part of popular culture such as time tunnels, teleportation, secret cities in Antarctica and reptilian races.

The big question is whether Carl Allen had read about these portals through Borderland Research or seen the article in *Life* about the Hungarian time ray experiments. He may have read the manuscript itself at some point as he was intensely interested in these subjects, but this would have been after 1960.

"The Hefferlin Manuscript" shows us how the "true" tales in *Amazing Stories* and the Shaver Mystery, which featured all of this stuff, were a potent and powerful influence on readers of the 1940s—and there were a lot of them. That included Carl Allen who must have been a reader of the popular magazine.

Carl Allen in his Later Years

Over the years various writers and researchers who tried to get more information from Carl Allen found his responses elusive, or could not find him at all.

In the 1960s it is said that several "fake" Allendes turned up offering to sell the whole story if the price was right. By this time the story had spread to the "fringe science" and "ufology" groups; the sudden appearance of all the fake Allendes caused some to doubt if there even was a real person named Carlos Allende.

Some people were apparently making money off the Allen/Allende story and as the years passed Allen became bitter at the unwanted publicity of his letters, and the simple fact that everyone but him was profiting from his story.

In 1969 Carl Allen/Allende visited the APRO (Aerial Phenomenon Research Organization) offices in Tucson, Arizona. At this time Allen handed a copy of the *Varo Edition* to Jim Lorenzen of APRO.

The following "confession" was attached to the second page of the Appendix:

> All words, phrases, and sentences underlined on the following pages in brown ink are false. The below page and the top part of the following were and are the craziest pack of lies I ever wrote. Object? To encourage ONR research and to discourage Professor Morris K. Jessup from going further with investigations possibly leading to actual research. Then I feared invisibility and force-field research; I don't now.

So, Carl Allen was essentially confessing that he was all three of the writers in the annotations to the Varo Edition.

Jim Lorenzen said that Allen had written the letters to Jessup

Carl Allen at APRO in 1969.

Carl Allen in a photo from Gray Barker, c.1978.

to "to scare the hell out of Jessup." Lorenzen also said:

> Allende still believes… that a US vessel… disappeared
> from its Philadelphia dock and reappeared seconds later
> in the Norfolk-Portsmouth area, then disappeared again to
> appear once more in its original berth.

Lorenzen would then publish an article in the APRO Bulletin
entitled: "Allende Letters a Hoax." (July–August 1969. pp. 1, 3.)

In 1980 Robert A. Goerman did some research into the
Philadelphia Experiment for an article he was writing for *FATE*
magazine. Goerman's conclusion after talking to Harold Allen
(Carl's father) and looking over some family documents was
that: "Carlos loves to play games with those foolish enough to
play audience…" In a letter to his parents, Carl admits annotating
Jessup's book, by himself.

Allen also wrote to Jacques Vallee between 1967 and 1968.
Allen offered to sell his acquired information, including a
special copy of the *Varo Edition* (which he had since entered

more notes into) for the "reduced" sum of $1,950. The first letter arrived in October 1967 from Monterrey, Mexico, the second was postmarked Minneapolis. Here is his statement to Dr. Vallee: "After this tremendous explosion my hair came out in bunches in my comb and I observed the same result in other Deck-crew members..."

Carl Allen in Colorado c.1985.

The above statement is interesting, and certainly resembles the effects of poisoning by nuclear radiation. Yet we have no evidence that Allen was ever exposed to any dose of radiation massive enough to cause his hair to fall out (he certainly seems to have grown it back to a normal pattern for an older man by the time photographs of him began to appear in the late 1980s), nor has anyone said he needed prolonged treatment, such as chemotherapy, for any later development of cancer.

To continue to quote Carl Allen: "I Watched it, saw it, observed its birth, growth, action and reaction upon the vehicle to which the super-field was being applied; I smelled it... my mouth tasted the ozone odor of it and my ears heard the sizzling-hum of its surrounding electrical envelope."

In 1977 Carl Allen made this statement to Bill Moore:

> I actually shoved my hand, up to the elbow, into this unique force field as that field flowed, surging powerfully in a counter clockwise direction around the little experimental Navy ship, the DE 173. I felt the...push of that force field against the solidness of my arm and hand outstretched into its humming-pushing-propelling flow.
>
> I watched the air all around the ship... turn slightly, ever so slightly, darker than all the other air... I saw, after a few minutes, a foggy green mist arise like a thin

cloud. This must have been a mist of atomic particles. I watched as thereafter the DE 173 became rapidly invisible to human eyes. ...in trying to describe the sounds that force field made as it circled around the DE 173... it began as a humming sound, quickly built up... to a humming whispering sound, and then increased to a strongly sizzling buzz—rushing torrent... The field had a sheet of pure electricity around it as it flowed. ...flow was strong enough to almost knock me completely off balance and had my entire body been within that field, the flow would of a most absolute certainty knocked me flat... on my own ship's deck. As it was, my entire body was not within that force field when it reached maximum strength—density... I was not knocked down but my arm and hand only pushed backward with the field's flow. ...Naval ONR scientists today do not yet understand what took place that day. They say the field was 'reversed'. Scientific history, I later came to realize, was made for the first time that day.

What occurred next? What did I observe next? I looked at the thin air where once had lain, and I could see, from the hull impression [in the ocean], the impression of the weight of the entire ship upon its bottom hull—that this impression, this weight, was causing an impression to be pushed, still, yet, down into the water as the DE sailed along, quite invisible, not visible to the human eye—I repeat, human eye—but there it sailed along making that impression, and that was the only evidence there was a solid, heavy, ship-shaped object of, at least, extreme transparency if not absolutely and totally invisible laying in that water propelling itself along at a normal speed, and then it scooted off and resumed patrol.

Sodium Pentothal—Truth Serum and Project Chatter

Sodium pentothal was discovered in 1922 but was not well known for decades. The Navy started to investigate its uses in 1947, so I just wonder how common knowledge it was in 1956 when Carl Allen wrote his letters to Jessup. Did Allen have some firsthand knowledge of these substances from Navy run experiments, or only "read" about them somewhere?

In 1947, the US Navy launched Project Chatter (1947–1953) and tested drugs for interrogation, which included experiments with mescaline, a hallucinogenic drug derived from the peyote cactus (with effects similar to LSD). Mescaline was studied as a possible speech-inducing agent after the Navy learned that Nazi doctors at the Dachau concentration camp had used it in mind control experiments. Years later in a written interview, Carl Allen was asked, "Would you still be willing to be hypnotized to bring out buried information from your subconscious about: The Invisibility Experiment, Your meeting with Einstein and Jessup, etc..." Carl Responded with three written words all in caps: "NOT WILLING, NO." We must consider that Carl Allen was used in mind control experiments, as he himself mentions it.

There is a good timeline of pharmaceuticals and government projects, including mind control, on Erowid.com:

1916	•Scopolamine is observed to promote unguarded speech by obstetrician Robert E. House. It will eventually be explored as a possible truth serum.
1931	•Research on the barbiturate sodium pentothal (the proverbial "truth serum") begins.
1942	•Franklin Delano Roosevelt orders the creation of a US intelligence service, designated the Office of Strategic Services (OSS), under the direction of General William "Wild Bill" Donovan.
ca. 1942 -43	•The OSS "truth drug" committee experiments with barbiturates, scopolamine, and Cannabis indica as possible truth serums.
1942	•British Special Operations Executive (SOE) begins Project SACCHARINE, investigating the use of various drugs to aid troops in combat. They evaluate several amphetamines as sources of emergency energy.

Oct 1942	•The OSS is charged with investigating the use of truth drugs to interrogate prisoners of war.
ca. 1943	•Nazi doctors experiment with mescaline as a truth serum at Dachau and Auschwitz.
Jan 30, 1943	•OSS unsuccessfully tests a mescaline and scopolamine cocktail as a truth drug on two volunteers at St. Elizabeth's Hospital.
May 1943	•OSS unsuccessfully tests THC as a truth serum.
1945	•The French Medico-Legal Society determines that confessions extracted under sodium pentothal are too unreliable to be used as evidence in court. Its conclusion was accepted by legal systems around the world.
Sep 20, 1945	•OSS is disbanded by Harry Truman. Planning immediately commences for the formation of a non-wartime intelligence service.
1946	•The Nuremberg Code is written after Nazi experiments on concentration camp inmates come to light during the Nuremberg Trials. The Code states that researchers must obtain full voluntary consent from all subjects, which is official US policy to this day.
1946	•US military intelligence concludes after exploring a number of compounds that the most effective truth drugs available are cannabis, followed closely by a combination of alcohol and caffeine.
Jan 1946	•The Central Intelligence Group, a precursor to the CIA, is founded.
1947	•The National Security Act creates the CIA and the National Security Structure.
1947	•The US Navy begins Project CHATTER, a "truth extraction method" research program. The project begins investigating possible truth serums.

1949	•Hungary tries Cardinal Josef Mindszenty, who publicly confesses to crimes he clearly did not commit. The CIA fears that he has been brainwashed by unknown means, reigniting US intelligence interest in mind control.
early 1950s	•The US Army Edgewood Arsenal begins investigating MDA (designated EA 1298) and MDMA (EA 1475) for possible use as interrogation tools. Toxicity studies are conducted on animals.
1950	•The term "brainwashing" appears for the first time in a Miami news article written by Edward Hunter, a CIA covert propaganda operator.
Apr 20, 1950	•CIA director Richard Hillenkoetter authorizes CIA Project BLUEBIRD, charged to investigate through scientific means various forms of mind control including interrogation techniques, brainwashing, and other behavioral research. The Nazi Dachau experiments are scrutinized, but are determined to be too saturated with sadism to be useful.
Jul 1950	•Project BLUEBIRD operatives experiment with interrogation techniques on a suspected double agent in Japan. They investigate debilitating heat, combinations of benzedrine and sodium amytal, and picrotoxin. Details are scarce, but the operation is considered a success.
Dec 1950	•Project BLUEBIRD begins experiments in applying electricity to the brain to effect mind control.

late 1950		•Psychology professor G. Richard Wendt receives $300K from Project CHATTER, along with 30 grams of pure heroin and 11 pounds of cannabis from the Federal Bureau of Narcotics. Student volunteers are administered drugs in double-blind studies. Volunteers are never told what they had been given.
1951		•Sidney Gottlieb becomes director of the CIA Technical Services Staff.
late 1951		•Project BLUEBIRD concludes that electroshock treatments can produce amnesia, though it also causes excruciating pain and sometimes reduces subject to a vegetative state.
1951		•Project BLUEBIRD is renamed Project ARTICHOKE and becomes a joint operation between the CIA and the US Army, Navy, and Air Force.
1952		•Project ARTICHOKE investigates neurosurgery as a tool for behavior modification.
1952		•Robert Hyde at Boston Psychopathic Hospital begins overseeing $40K a year in CIA funds for LSD research.
spring 1952		•Harvard professor Henry Beecher alerts the British Joint Intelligence Bureau (JIB) to LSD.
mid-1952		•CIA Operation CASTIGATE, directed by US Navy Project CHATTER personnel, conducts experiments on a truth serum in Germany. The operation is an embarrassing failure for the Navy, and Project CHATTER never recovers. The CIA now has primary ownership of mind control research.
1952		•Morse Allen, head of Project ARTICHOKE, hears rumors of psychoactive mushrooms, and begins the CIA search for them.

Apr 13, 1953		•CIA director Allen Dulles authorizes Project MKULTRA under the direction of Sidney Gottlieb. Gottlieb later testifies that MKULTRA's mission was "to investigate whether and how it was possible to modify an individual's behavior by covert means."
1953		•MKULTRA focuses its attention strongly on LSD.
May 1953		•MKULTRA hires former Narcotics Bureau officer George White to run a safe house in New York City. In an operation that will come to be called Midnight Climax, the safe house is used to surreptitiously slip drugs to civilians so their responses can be monitored.
1953		•Under the direction of MKULTRA, Dr. Harris Isbell, director of the Addiction Research Center, begins performing drug tests on his inmate population.
1953		•The Sandoz patent on LSD expires, allowing US companies to legally manufacture LSD for the CIA for the first time.
Sep 1953		•The CIA negotiates a deal with Sandoz not to sell LSD to the Soviet Union.
late 1953		•CIA operatives regularly dose one another unawares with LSD as an ongoing operational experiment. Some agents have extremely negative reactions.
late 1953		•British intelligence administers LSD to Royal Air Force volunteers. They eventually conclude that 'Research is desirable into the use of LSD-25 as a possible effective agent for use in interrogation.'

late 1953	•CIA director Dulles commissions Dr. Harold Wolff to conduct a study on communist brainwashing techniques. After conducting an exhaustive study, Wolff concludes that the Chinese and Soviets use a combination of coercion, stress, and pressure, and have no machines, pills, or rays.
Nov 19, 1953	•Sidney Gottlieb doses Frank Olson and several other US Army Special Operations Division officers at a CIA/Army joint retreat. Olson responds badly, and falls into lasting depression, vacillating between apparent normalcy and severe paranoia and depression.
Nov 28, 1953	•Frank Olson falls from a hotel window to his death. The event is described as a suicide, though serious questions are later raised about this explanation.
1954	•US pharmaceutical company Eli Lilly develops an entirely-synthetic fabrication procedure for LSD. No longer constrained by limited ergot supplies, Eli Lilly produces LSD in quantity for the CIA.
1955	•CIA agent Morse Allen and Dr. Maitland Baldwin of the NIH propose extreme sensory deprivation experiments. The proposal is shot down during review by CIA medical agents, who suggest that the experimenters consider "volunteer[ing] their heads for use in Dr. Baldwin's 'noble' project."
early 1955	•George White opens additional safe houses in San Francisco and Marin, and oversees them along with the New York safe house. Prostitutes are paid to surreptitiously dose their customers with LSD, while being monitored by CIA operatives. This operation is now referred to as Operation Midnight Climax.

46

Jun 29, 1955		•R. Gordon Wasson participates in a psilocybin mushroom velada in Mexico. A few months later he is contacted by Professor James Moore, who asks to accompany Wasson on his next expedition the following summer, and offers to underwrite the trip with a $2K grant from the Geschickter Fund. Unbeknownst to Wasson, Moore is a CIA agent, and the Geschickter Fund is an MKULTRA conduit.
1956		•Dr. Ewen Cameron tests LSD in conjunction with "depatterning" experiments designed to reprogram personalities. His research soon comes to the attention of the CIA.
mid-1956		•CIA operative James Moore accompanies Wasson and Robert Heim to Mexico and brings back samples of psilocybin mushrooms. CIA agents are unable to isolate the active principle.
1957		•A CIA report states that six psychoactive drugs have been moved out of the experimental stage and into operational use.
1957		•The CIA begins issuing sizable grants to Dr. Ewen Cameron to support his "depatterning" research. Cameron also begins studying sensory deprivation in this year.
Nov 1957		•The CIA Intelligence Center at Fort Holabird and the Chemical Warfare Laboratories at Edgewood begin working together. Thirty to thirty-five volunteers are administered LSD, some of them as many as twenty times in a two-year period.
1958		•John Lilly resigns from NIH, in part because of fears that the military will use his consciousness research in unethical ways.

1958		•Under US Army contract, Dr. Gerald Klee administers LSD to volunteers in doses as large as 1200 ug.
1959		•Albert Hofmann publishes the synthesis of psilocybin, which the CIA has been unable to identify. Dr. Harry Isbell soon procures a quantity and orders it injected into nine inmates at his facility.
late 1950s		•US Army doctor Van Sim and his colleagues covertly dose Edgewood Arsenal volunteers with LSD.
late 1950s		•Five civilian volunteers at Edgewood Arsenal are given PCP in a search for incapacitating agents. Experiments are discontinued after one subject ends up in the hospital for six weeks with paranoid psychosis.
Apr 15, 1961		•Bay of Pigs invasion ends in disaster. President Kennedy vows to "splinter the CIA into a thousand pieces." CIA Director Dulles and Deputy CIA Director Charles Cabell are forced to resign. John McCrone is appointed the new Director and is instructed to clean house.
1963		•Newly-appointed CIA Inspector General John Earman urges CIA director John McCrone to close the Operation Midnight Climax safe houses.
1964		•Project MKULTRA becomes Project MKSEARCH.
1965		•The CIA closes the San Francisco Midnight Climax safe house.
1966		•The CIA closes the New York City Midnight Climax safe house.
1966		•The CIA continues drug-assisted interrogations until at least this year.

1966		•MKSEARCH funds the Amazon Natural Drug Company to investigate plant-based toxins and drugs. The group researches yage/ayahuasca and numerous other psychoactive substances.
1967		•The Geschickter Fund is discontinued as a CIA conduit.
1968		•CIA Project OFTEN begins at the US Army Edgewood Arsenal to investigate the effects of various drugs on animals and humans.
1972		•Gottlieb ends MKSEARCH, observing "It has become increasingly obvious over the last several years that this general area has less and less relevance to current clandestine operations."
1972		•Richard Nixon purges CIA director Richard Helms for reasons that are unclear. James Schlesinger is named head of the Agency.
1972		•Helms and Gottlieb order destruction of all MKULTRA records. As a result of a clerical error, seven boxes of documents are spared.
1972		•In the aftermath of Watergate, CIA director Schlesinger orders all CIA employees to inform him of illegal actions they have conducted in their operations, and he learns of Frank Olson's death. Part of the resulting report is leaked to investigative journalist Seymour Hersh.
1973		•Project OFTEN is canceled.
May 1974		•Seymour Hersh writes an article for the New York Times detailing criminal CIA domestic operations, creating an uproar. President Ford appoints committee chaired by VP Nelson Rockefeller to investigate intelligence improprieties.

Project Rainbow and the Philadelphia Experiment

Tuesday, March 8, 1994

A4 GREELEY (Colo.) TRIBUNE

FOR THE RECORD

OBITUARIES

Carlos Allende

Carlos M. Allende, 68, Greeley, died Saturday, March 5, at Centennial Health Care Center.

He was born May 31, 1925, in Springdale, Pa., to Harry C. and Mary M. (Mahaffey) Allen.

Mr. Allende moved with his family to Lower Burrell, Pa., when he was 15 years old.

He joined the U.S. Marine Corps during World War II, and then the Merchant Marines. He lived through the United States, in Colorado and Mexico and came to Greeley in 1986.

Survivors are two brothers, David R. Allen of Lower Burrell and Richard Allen of Natrona Heights, Pa. A sister, Elsie Schearer, and a brother, Fred Allen, are deceased.

Services will be at 10 a.m. Thursday at Stoddard Funeral Home. Interment will be in Sunset Memorial Gardens.

Memorial gifts may be made to Resident Activity and Peoples Library at Centennial Health Care Center in care of Stoddard Funeral Home.

Carlos Allende/Carl Allen's obituary. in the *Greeley Tribune*.

Chapter 3

The US Navy and the Varo Edition

I don't ever want to play the part
Of a statistic on a government chart
There has to be an invisible sun
—*Invisible Sun*, The Police

As mentioned in the last chapter, in late 1955 Carl Allen sent an anonymous package marked "Happy Easter" containing a copy of Morris K. Jessup's book *The Case for the UFO: Unidentified Flying Objects* to the US Office of Naval Research (ONR). The book was filled with handwritten notes in its margins, written with three different shades of blue ink, appearing to detail a debate among three individuals. Only one was given a name and that name was "Jemi."

The three different scripts commented on Jessup's ideas about the propulsion of flying saucers, such as anti-gravity, discussed alien races, and expressed concern that Jessup was too close to discovering the technology which was presumably from outer space.

The commenters referred to each other as "gypsies," and discussed two different types of "people" living in outer space. Their text contained non-standard use of capitalization and punctuation, and detailed a lengthy discussion of the merits of various elements of Jessup's assumptions and speculations in the book. There were oblique references to the Philadelphia Experiment (one commenter reassures his fellow annotators who have highlighted a certain theory which Jessup advanced).

As related in the previous chapter, shortly after sending the package to the ONR in January, 1956, Allen began sending a series of letters to Jessup, using his given name as well as "Carlos Miguel Allende." The first known letter warned Jessup not to investigate the levitation of unidentified flying objects. Allen put forward a story of dangerous science based on alleged unpublished theories by Albert Einstein. He further claimed a scientist named Franklin Reno put these theories into practice at the Philadelphia Naval Shipyard in October 1943.

When searching for "Franklin Reno" on the Internet one gets a University of Michigan archived set of documents of a 1937 to 1943 investigation of a "Franklin Victor Reno" by the Military Intelligence Division of the War Department.

This 9-page document (plus some others) says that he was being investigated because he was "a leader of the National Student League and was suspected of being a communist." The document also says that he worked as a mathematician at the Ballistic Laboratory at the Aberdeen Proving Ground in Maryland. The investigation ended in 1943. Whether this is the scientist that Allen was referring to cannot be known, but it is possible that Allen met Reno during 1943 or in the years after.

Allen claimed in the letters to have witnessed this experiment while serving aboard the SS *Andrew Furuseth*. In his account, a destroyer escort was successfully made invisible, but the ship inexplicably teleported to Norfolk, Virginia for several minutes, and then reappeared in the Philadelphia yard. The ship's crew was supposed to have suffered various side effects, including insanity, intangibility, and being "frozen" in place. When Jessup wrote back requesting more information to corroborate his story Allen said his memory would have to be recovered and referred Jessup to the Philadelphia newspaper article that Allen claimed covered the incident and was actually about "disappearing" sailors in a bar near the docks.

In 1957 the ONR invited Jessup to visit its offices where he was shown the annotated copy of his book. Jessup noticed the handwriting of the annotations resembled the letters he had received from Allen.

Two officers at ONR, Captain Sidney Sherby and Commander

George W. Hoover, took a personal interest in the matter. Hoover later explained that his duties as Special Projects Officer required him to investigate many publications and that he ultimately found nothing of substance to the alleged invisibility experiment. Hoover discussed the annotations with Austin N. Stanton, president of Varo Manufacturing Corporation of Garland, Texas during meetings about Varo's contract work for ONR.

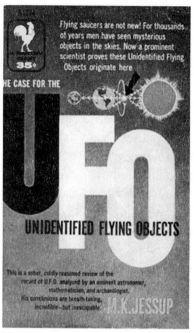

The paperback edition of Jessup's book.

According to an Internet search, the Varo Manufacturing Corporation is located in Garland, Texas. The company is categorized as Power Factor Correction Manufacturers. Records say that it was established in 1946 and incorporated in Texas. The company still exists today and has an annual revenue of $37,644,317.00 and employs a staff of approximately 412.

Austin Stanton of Varo became so interested that Varo's office began producing mimeographed copies of Jessup's book with the annotations and Allen's letters, first a dozen and eventually 127 copies. These copies came to be known as the "Varo Edition." Besides noting handwriting of the individual named "Jemi" (addressed as such by the others and using blue-violet ink), the anonymous introduction (done by the Varo stenographer Michael Ann Dunn) to the Varo Edition concludes that there were two other individuals making annotations, "Mr. A" (identified as Allen by Jessup, in blue ink), and "Mr. B" (in blue-green ink).

The big mystery about the Varo Edition is why did a government contractor go through so much trouble to reprint a book that had been rejected by the scientific community? Why were the annotations in the book so interesting to the Navy and Austin Stanton?

It appears that the book may have been printed by the Varo company at the request of the military which implies that the government was very interested in the comments being made by the three "gypsies" in the margins of the book. Why were the Navy and others so fascinated by these comments? Their interest in the bizarre comments and subject matter in the book tends to give the annotations some credence.

Was the Navy trying to figure out for themselves if some of comments were true? Had a ship been turned invisible while at the Philadelphia Naval Yard? Had sailors been fused to parts of the ship or otherwise lost their minds? Was the Navy essentially investigating itself? If parts of the comments were in fact true, did the Navy realize that some of their secrets were being told to Jessup? Jessup's mysterious death, to be discussed in the next chapter, would only add to the suspicions that the Navy was intensely interested in the subject and the people who were playing a part in it, including Carl Allen and Morris Jessup.

In July of 1973 Grey Barker, the publisher of Saucerian Press in West Virginia, published a facsimile edition of the Varo Edition to Jessup's *The Case for the UFO*. In it he included an introduction written by Michael Ann Dunn the stenographer who did the writing. In it she said:

> It appears that these notes were written by three persons. The use of three distinct colors of ink—blue, blue-violet, and blue-green—and the difference in handwriting lead to this conclusion. Hereafter they will be referred to as Mr. A, Mr. B, and Jemi.
>
> It is assumed that the third person was Jemi because of the direct use of "Jemi" in salutations and references to that name by Mr. A and Mr. B throughout the book. There are many, some of which appear on pages 2, 81, 122, 126, and 162 in the original book. It is possible of course that it is merely a salutation.
>
> It is possible that two of these men are twins. There are two references to this word. They appear on page 6 and page 81 of the original book. The assumption that Mr. A is one of the twins may be correct. On page 81, Mr. A

has written and marked through "…and I Do Not know How this came to Pass, Jemi." Then he has written, "I remember, My twin…" On page 6 he writes in an apparent answer to Mr. B, "No, My twin…" We cannot be sure of the other twin.

It is probable that these men are Gypsies. In the closing pages of the book Mr. B says, "…only a Gypsy will tell another of that catastrophe. And we are a discredited people, ages ago. Hah! Yet, man wonders where "we" come from…" On page 130 Mr. A. says, "…ours is a way of life, time proven & happy. We have nothing, own nothing except our music & philosophy & are happy." On page 76 Mr. B. says "Show this to a Brother Gypsy…" On Page 158 the reference to the word "we" by Mr. A could refer to the "discredited people." Charles G. Leland in his book "English Gipsies and Their Language" states that the Gypsies call each other brother and sister, and are not in the habit of admitting to their fellowship people of a different blood and with whom they have no sympathy. This could explain the usage of the term in the closing notes "My Dear Bothers" and perhaps the repeated reference to "vain humankind."

This book was apparently passed through the hands of these men several or many times. This conclusion is drawn from the fact that there are discussions between two or all three of the men, questions answered, and places where parts of a note have been marked through, underlined, or added to by one or both of the other men. Some have been deleted by marking through.

…

It might be helpful for you to know a little about the nature of the notes before you begin reading this book. The notes refer to two types of people living in space. Specifically the "stasis neutral" and the undersea are mentioned as habitats. They seem to live in both interchangeably. The building of undersea cities is mentioned. Many different kinds of ships are used as transportation. These two peoples, races or whatever they may be called, are referred to over and over again. They are called L-M's and S-Ms. The L-M's seem to be peaceful; the S-M's are not. It seems that the annotations are inclined toward the L-M's as they speak more kindly of them that the S-M's.

Terms such as: mothership, home-ship, dead-ship, Great ark, great bombardment, great return, great war, little-men, force-fields, deep freezes, undersea building, measure markers, scout ships, magnetic and gravity fields, sheets of diamond, cosmic rays, force cutters, undersea explorers, inlay work, clear-talk, telepathing, burning "coat", nodes, vortice, magnetic "net", and many others are used quite naturally by these men. They explain how, why, and what happens to people, ships, and planes that have disappeared. They explain the origin of odd storms and clouds, objects falling from the sky, strange marks and footprints, and other things which we have not solved.

We now know that it is likely that Carl Allen was all three of the "gypsies" that were writing the notes. Allen seems to have considered himself a gypsy as he was continually traveling around the United States and Mexico and did not have a permanent job or profession. It is unknown how Allen earned a living during all of his years, presumably he was doing odd jobs at various locations.

Jacques Vallée's Research into Project Rainbow

Researcher Jacques Vallée describes a procedure on board USS *Engstrom*, which was docked alongside the *Eldridge* in 1943. The operation involved the generation of a powerful electromagnetic field on board the ship in order to deperm or degauss it, with the goal of rendering the ship undetectable or "invisible" to magnetically fused undersea mines and torpedoes.

This system was invented by a Canadian, Charles F. Goodeve, when he held the rank of commander in the Royal Canadian Naval Volunteer Reserve, and the Royal Navy and other navies used it widely during World War II.

Sir Charles Frederick Goodeve OBE FRS (born February 21, 1904—died April 7, 1980) was a Canadian chemist and pioneer in operations research. During World War II, he was instrumental

USS *Engstrom* underway off the Mare Island Naval Shipyard February 27, 1945.

in developing the "hedgehog" antisubmarine warfare weapon and the degaussing method for protecting ships from naval mines.

Goodeve had served in the Royal Canadian Naval Volunteer Reserve on the HMCS *Chippawa*. In England, he joined the Royal Naval Volunteer Reserve. In 1936, he was promoted to the rank of Lieutenant Commander. In 1939, he began work at HMS *Vernon*, specializing in ways to counter the threat of mines. He developed the "Double L" technique for minesweeping magnetic mines. Later he developed the degaussing method for reducing the magnetic field around ships that triggered mines. Goodeve coined the term after the gauss unit, used by the Germans during the war to measure magnetic fields that they named after German scientist Carl Friedrich Gauss (1777–1855). Goodeve also developed the related technique of "wiping."

In 1940, Goodeve implemented the British production of the Swiss-designed Oerlikon 20 mm cannon, which was needed as anti-aircraft protection on naval and merchant ships. His group, renamed the Directorate of Miscellaneous Weapons Development (DMWD), then worked on antisubmarine warfare developing the hedgehog, an array of spigot mortars which threw small, contact-

Carl Friedrich Gauss.

fused bombs ahead of a ship. At one point, to protect the project from internecine warfare inside the Royal Navy, Goodeve finagled a demonstration of a prototype for Prime Minister Winston Churchill. For his work in weapon development, Goodeve was awarded an OBE (Order of the British Empire).

In 1942, the Third Sea Lord, Vice Admiral Sir William Wake-Walker, appointed Goodeve Assistant Controller Research and Development, with broad oversight of the Navy's research and development efforts. At the end of the war, Goodeve was knighted, and awarded the US Medal of Freedom. According to Wikipedia:

Degaussing is the process of decreasing or eliminating a remnant magnetic field. It is named after the gauss, a unit of magnetism, which in turn was named after Carl Friedrich Gauss. Due to magnetic hysteresis, it is generally not possible to reduce a magnetic field completely to zero, so degaussing typically induces a very small "known" field referred to as bias. Degaussing was originally applied to reduce ships' magnetic signatures during World War II. Degaussing is also used to reduce magnetic fields in cathode ray tube monitors and to destroy data held on magnetic storage.

The mines detected the increase in the magnetic field when the steel in a ship concentrated the Earth's magnetic field over it. Admiralty scientists, including Goodeve, developed a number of systems to induce a small "N-pole up" field into the ship to offset this effect, meaning that the net field was the same as the background. Since the Germans used the gauss as the unit of the strength of the magnetic field in their mines' triggers (not yet a standard measure), Goodeve referred to the various processes to counter the mines as "degaussing." The term became a common word.

The original method of degaussing was to install electromagnetic coils around the ships, known as coiling. In addition to being able to bias the ship continually, coiling also allowed the bias field to be reversed in the southern hemisphere, where the mines were set to detect "S-pole down" fields. Most British ships, notably cruisers and battleships, were well protected by about 1943.

Installing such special equipment was, however, far too expensive and difficult to service all ships that would need it, so the navy developed an alternative called wiping, which Goodeve also devised. In this procedure a large electrical cable was dragged upwards on the side of the ship, starting at the waterline, with a pulse of about 2000 amperes flowing through it. For submarines, the current came from the vessels' own propulsion batteries. This induced the proper field into the ship in the form of a slight bias.

It was originally thought that the pounding of the sea and the ship's engines would slowly randomize this field, but in testing, this was found not to be a real problem. A more serious problem was later realized: as a ship travels through Earth's magnetic field, it will slowly pick up that field, counteracting the effects of the degaussing. From then on captains were instructed to change direction as often as possible to avoid this problem. Nevertheless, the bias did wear off eventually, and ships had to be degaussed on a schedule. Smaller ships continued to use wiping through the war.

To aid the Dunkirk evacuation, the British "wiped" 400 ships in four days. Also during World War II, the United States Navy commissioned a specialized class of degaussing ships that were capable of performing this function. One of them, USS *Deperm*, was named after the procedure.

After the war, the capabilities of the magnetic fuzes were greatly improved, by detecting not the field itself, but changes in it. This meant a degaussed ship with a magnetic "hot spot" would

RMS *Queen Mary* arriving in New York Harbor, June 20, 1945, with thousands of US soldiers—note the prominent degaussing coil running around the hull.

still set off the mine. Additionally, the precise orientation of the field was also measured, something a simple bias field could not remove, at least for all points on the ship. Therefore a series of ever-increasingly complex coils were introduced to offset these effects, with modern systems including no fewer than three separate sets of coils to reduce the magnetic field in all axes.

In April 2009 the US Navy tested a prototype of its High-Temperature Superconducting Degaussing Coil System, referred to as "HTS Degaussing." The system works by encircling the vessel with superconducting ceramic cables whose purpose is to neutralize the ship's magnetic signature, as in the legacy copper systems. The main advantage of the HTS Degaussing Coil system is greatly reduced weight—sometimes by as much as 80%—and increased efficiency. According to Wikipedia:

> A ferrous-metal-hulled ship or submarine, by its very nature, develops a magnetic signature as it travels, due to a magneto-mechanical interaction with Earth's magnetic field. It also picks up the magnetic orientation of the Earth's magnetic field where it is built. This signature can be exploited by magnetic mines or facilitate the detection of a submarine by ships or aircraft with magnetic anomaly detection (MAD) equipment. Navies use the deperming procedure, in conjunction with degaussing, as a countermeasure against this.

> Specialized deperming facilities, such as the United States Navy's Lambert's Point Deperming Station at Naval Station Norfolk, or Pacific Fleet Submarine Drive-In Magnetic Silencing Facility (MSF) at Joint Base Pearl Harbor–Hickam, are used to perform the procedure.

> During a close-wrap magnetic treatment, heavy-gauge copper cables encircle the hull and superstructure of the vessel, and high electrical currents (up to 4000 amperes) are pulsed through the cables. This has the effect of "resetting" the ship's magnetic signature to the ambient level after flashing its hull with electricity. It is also possible to assign a specific signature that is best suited to the particular area of the world in which the ship will

operate. In drive-in magnetic silencing facilities, all cables are either hung above, below and on the sides, or concealed within the structural elements of facilities. Deperming is "permanent." It is only done once unless major repairs or structural modifications are done to the ship.

Vallée speculates that accounts of USS *Engstrom*'s degaussing might have been garbled and confabulated in subsequent retellings, and that these accounts may have influenced the story of "The Philadelphia Experiment."

Vallée cites a veteran who served on board USS *Engstrom* and who suggests it might have traveled from Philadelphia to Norfolk and back again in a single day at a time when merchant ships could not, by use of the Chesapeake & Delaware Canal and Chesapeake Bay, which at the time was open only to naval vessels. Use of that channel was kept quiet: German submarines had ravaged shipping along the East Coast during Operation Drumbeat, and thus military ships unable to protect themselves were secretly moved via canals to avoid the threat.

The same veteran claims to be the man that Allende read about in the newspaper article "disappearing" at a bar. He claims that

Close-wrap deperming of Ivan Gren-class landing ship, 2016.

when a fight broke out, friendly barmaids whisked him out of the bar before the police arrived, because he was under the legal age for drinking. They then covered for him by claiming that he had disappeared.

So, Vallée believes that Allen witnessed the degaussing of the *Eldridge* and believed that it had been an invisibility experiment. When Carl Allen visited APRO they thought that Allen actually believed in the invisibility experiment even though he admitted to writing all the notes in the copy of *The Case for the UFO* sent to the ONR.

Was he just making up all this stuff or did he really believe these things had happened? This is the problem with his annotated copy of Jessup's book and the letters he wrote to Jessup.

Jessup Visits with the Navy

Riley Crabb, the former director of the Borderland Research Sciences Foundation in California, was one of the first people to write about the Philadelphia Experiment. He published a mimeographed booklet in 1962 entitled *M.K. Jessup, the Allende Letters and Gravity.* According to Crabb, an annotated copy of *The Case for the UFO* was addressed to Admiral N. Furth, Chief, Office of Naval Research, Washington 25, D.C., and was mailed in a manila envelope postmarked Seminole, Texas in 1955. In July or August of that year the book appeared in the incoming correspondence of Major Darrell L. Ritter, USMC, Aeronautical Project Office in ONR. When Captain Sidney Sherby, a newcomer, reported aboard at ONR he obtained the book from Major Ritter. Captain Sherby and Commander George W. Hoover, Special Projects Officer, ONR, indicated interest in some of the notations the book contained.

As noted earlier, the paperback copy of Jessup's book was marked throughout with underlinings and notations, evidently by three different persons, for three distinct colors of ink were used—blue, blue-violet, and blue-green. The notations in Jessup's book implied intimate knowledge of flying saucers, their means of propulsion, their origin, background, history, and even the habits of the beings occupying them. In the annotated bits in the margins of the book were references and unusual terms such as mothership, deep freezes, sheets of diamond, and other strange references as listed by Michael Ann Dunn.

Hoover and Sherby invited Jessup to visit them at the Office of Naval Research in Washington in the spring of 1957 to discuss the book they had received. They handed Jessup the annotated copy and asked him to examine the handwritten notes.

"This book was sent to us through the mail anonymously. Apparently it was passed back and forth among at least three persons who made notations. Look it over, Mr. Jessup, and tell us if you have any idea who wrote those comments," one of the officers said.

As Jessup looked at the book, he was troubled at what he saw. Due to the uniqueness of Allende's spelling and grammar, it became almost immediately evident to Jessup that the letter-writer Allende was possibly one of those who had worked on the annotation of his book, and was very likely the same individual who had originally mailed the volume to the Navy in the first place. The many references were startling and showed that the writer(s) were extremely knowledgeable in the subject of UFOs, folklore, mysticism and physics. Further troubling was the fact that Navy took such an interest in these strange ramblings. Jessup finally told them that he had received two letters from one of the writers in the book, Carlos Allende/Allen.

"Thank you, Mr. Jessup," he was told. "It is important that we see those letters." Hoover then told Jessup that he was arranging for the printing of a special edition of the book "for some of our top people," and that he would be sure to arrange for Jessup to get a copy.

Jessup returned to his home and supplied Hoover and Sherby with copies of the letters. The two officers then arranged for the Varo Manufacturing Company of Garland, Texas to produce a mimeographed edition of the book. Brad Steiger, a well-known author on occult and paranormal issues, claims that the Varo company is engaged in "secret government work." This special edition of Jessup's book, with the handwritten notes in the margins, became known as "the Varo annotated edition," or just "the Varo Edition."

Since the book allegedly involved the military and their

apparent knowledge of technical details concerning UFOs, the Varo Edition of *The Case for the UFO* caused considerable interest within the fringe community of researchers who were studying the UFO enigma.

Brad Steiger says that Riley Crabb sent correspondence to him on September 24, 1962, where Crabb clears up the mystery of how he knew of the Varo Edition's history, and how he happened to obtain a copy of the original Varo Edition. It was a copy that the Navy originally gave to Jessup, and apparently was given to Crabb by Jessup. This copy, however, rather mysteriously disappeared in April, 1960, when Crabb mailed it to himself from Washington.

In elaborating further about the original Varo edition, Crabb wrote to Steiger:

> It may be that CDR Hoover, or some other open minded officer in the Office of Navy Research asked Varo to print the notes and the book; on the other hand, maybe Bob Jordan or some other official at Varo, in Washington at the time, sniffed a possible research and development project, and volunteered to do it. Anyhow, I understand that 25 copies were reproduced on standard letter paper, 8 1/2 x 11, probably on Varo's own little litho press, and plastic bound, pretty close to 200 pages. Michael Ann Dunn, the stenographer who did the editing, explains why in the introduction. She's married now, living in Dallas, and won't answer her phone. Garland (where Varo is located) is a Dallas suburb. I suppose Varo and/or the Navy judiciously handed out a few copies to those capable of picking up a hot lead on the anti-gravity research trail. From their comment, Allende and his gypsy friends were not educated, not technicians, and did not give any illustrations or mathematical formulae which would help build usable hardware.
>
> Varo, by the way, is a small manufacturing firm in electronics and up to its neck in space age business. Apparently it has succeeded in developing some kind of a death ray gadget, judging from a guarded press release of last fall when a group of Congressmen visited there

for a demonstration. I think it extremely interesting that certain Naval officers and Varo officials took the Allende notes far more seriously at first reading than did Jessup himself! My first reaction to them was one of skepticism, but now I believe them. So does one of our most material minded, hardest headed electronic engineer associates in Los Angeles.[9]

The Varo Edition of Jessup's book may have alarmed certain commanders at the Office of Naval Research. Perhaps they thought that a very serious security leak was occurring, and Jessup had found out too much concerning the Navy's secret time travel and teleportation experiments.

Chapter 4

The Life and Death of
Morris K. Jessup

Can you feel it
Now that spring has come
And it's time to live in the scattered sun
—*Waiting for the Sun*, The Doors

Morris Ketchum Jessup was born in Rockville, Indiana on March 2, 1900. Though employed for most of his life as an automobile-parts salesman and a photographer, he is probably best remembered for his writings on UFOs.

Jessup grew up with an interest in astronomy. It is thought that he earned a Bachelor of Science degree in astronomy from The University of Michigan in Ann Arbor in 1925 and received a Master of Science degree in 1926. Though he began work on his doctorate in astrophysics, he ended his dissertation work in 1931 and never earned the higher degree. Nevertheless, he was sometimes referred to as "Dr. Jessup." He apparently dropped his career and studies in astronomy and worked for the rest of his life in a variety of jobs unrelated to science, although he is sometimes erroneously described as having been an instructor in astronomy and mathematics at the University of Michigan and Drake University.

Jessup has been referred to in ufological circles as "probably the most original extraterrestrial hypothesizer of the 1950s," and it has been said of him that he was

Jessup in a 1917 yearbook.

67

Morris Jessup

Morris Jessup in the Navy.

"educated in astronomy and archeology and had working experience in both."

Jessup took part in archeological expeditions to the Yucatan and Peru in the 1920s. In 1923, along with University of Michigan botanist Carl D. LaRue and plant pathologist James Weir, Jessup participated as a photographer in a US Department of Agriculture expedition to Brazil to study the possibility of growing rubber in the Amazon. Henry Ford would later draw on the expedition's findings, as well as the assistance of LaRue and Weir, when planning Fordlandia, his rubber plantation in the Amazon.

As a masters student in the late 1920s, he travelled to South Africa and worked at the Lamont-Hussey Observatory in Bloemfontein.

During these years there were few jobs for astronomers or astrophysicists, so he worked for the US Department of Agriculture as part of team of scientists going to Brazil to "study the sources of crude rubber in the headwaters of the Amazon." A strange assignment indeed, since the headwaters of the Amazon actually would be situated in Peru, Bolivia or Ecuador.

M. K. Jessup, Carl LaRue, and James Weir in Brazil on an expedition for the US Department of Agriculture, 1923.

Jessup's World Travels

Jessup, during these years, and on subsequent journeys to Mexico and Peru, visited the massive ruins at Teotihuacan, and at Cuzco, Sacsayhuaman, Ollantaytambo and Machu Picchu. Jessup concluded that these Peruvian giant walls could not have been built by the Incas, as archaeologists commonly suggest. Rather, he formulated one of the earliest "Gods from Outer Space" theories.

Jessup documented an expedition to Cuzco he took part in during 1930. In 1934 Jessup wrote an article for *American Anthropologist* titled "Inca Masonry at Cuszco" (*American Anthropologist*, Number 35, 1934, pp. 239-241). In the article he shares a photo of the famous stone of 12 angles in Cuzco and says:

> In October 1930 the writer had an opportunity to examine the Inca ruins at Cuzco during an all too brief visit to the historic city. One or two features of the ancient stone work which were noticed at that time seem worth recording.
>
> It has often been pointed out that the workmanship in Inca masonry is of a superb nature with joints so perfect as not to permit the penetration of a penknife. It is a further matter of note that each stone is so formed that it can occupy one and only one position in the walls, and that no mortar was used in the construction. It is the purpose of this paper to explain the first two points and show that mortar is unnecessary.
>
> Let us study Plate 5, A. This is the famous "stone with twelve corners," which is so often cited as an example of the careful forethought used in planning a wall so that each stone fits a certain niche. On a casual inspection of this stone it occurred to the writer that the stones were not quarried to these weird shapes, but were quarried roughly and then ground to the final shape in situ. The photos seem to prove this assumption.

So we see here the cautious early research into the amazing stonework seen at Cuzco and other so-called Inca sites nearby.

The famous stone of 12 angles that Jessup used in his article on Inca masonry.

Even today this style of masonry has not been adequately explained and in some cases we are talking about stones that weigh 125 tons or more. Many people, including myself, have concluded that the Incas did not build these monuments and they merely occupied the virtually indestructible stone buildings built by an earlier culture.

But who was this earlier society that built gigantic and perfect walls out of granite at some point in prehistory? Were they extraterrestrials? We see that as early as 1930 Jessup was leaning toward what we now call "Ancient Astronaut Theory."

Then came World War II with its mysterious "Foo Fighters" and the flying saucer sightings that began after the war. Jessup would not know anything about the many missing Nazi submarines, the secret German base in Antarctica, or how many thousands of Nazis had escaped Europe through Italy and Spain to find new lives in Argentina and other countries. Like most UFO researchers of the 1950s and 60s he would be unaware of the German creation of flying saucers such as the Vril and Haunebu craft or of the cylindrical mothership called the Andromeda. I detail these in four books published between 2020 and 2024 on the survival of Nazis and the SS in South America.[25, 26, 27, 28]

However, Jessup was a freethinker and already impressed by the inexplicable stonework he had seen in Peru. He was also interested in flying saucers and advanced technology. He put all of

70

INCA MASONRY AT CUZCO   *By* MORRIS K. JESSUI

IN OCTOBER 1930 the writer had an opportunity to examine the Inci
ruins at Cuzco during an all too brief visit to that historic city. One o
two features of the ancient stone work which were noticed at that time seen
worth recording.

It has often been pointed out that the workmanship in Inca masonr;
is of a superb nature with joints so perfect as not to permit the penetratioi
of a penknife. It is a further matter of note that each stone is so formei
that it can occupy one and only one position in the walls, and that no morta
was used in the construction. It is the purpose of this paper to explain thi
first two points and show that mortar is unnecessary.

Let us study Plate 5, A. This is the famous "stone with twelve corners,'
which is so often cited as an example of the careful forethought used ii
planning a wall so that each stone fits a certain niche. On a casual inspectioi
of this stone it occurred to the writer that the stones were not quarrie(
to these weird shapes, but were quarried roughly and then ground to thei
final shape *in situ*. The photos seem to prove this assumption.

In 5, A we may assume the lowest layer to be in place. The mason thei
lays the rough stone, No. 1, on the wall and his helpers proceed to fit i
by pulling it back and forth at right angles to the face of the wall until it
left hand and bottom surfaces fit perfectly against those of their neighbors
Large twelve-cornered stone, No. 2, was next in place. The curvature o
the joint, 1-2, and the rounded corners indicate grinding *in situ*. Thiri
in place was No. 3, again with a curved joint. Numbers 4, 5, 6, 7, 8, 9
were put on in order, each being perfectly fitted before the next was started
Notice how No. 4 was worked in as a wedge under its own weight. Th
rounded corners and curved sides are a natural result of grinding. Corner
4-5-6, 2-7-8, and 2-8-9 are especially interesting in this respect.

It is at once obvious that joints so made could easily reach a perfectioi
such that the penetration of a knife would be difficult or impossible, an(
expecially so if the fine sand and powder caused by the grinding were per
mitted to remain in the joints, which seems to have happened in some case:

Plate 5, B shows a wall of somewhat better grade; at least there is mor
attention to detail. Here there is ample material for checking the assump
tion that stones were fitted *in situ*. Without going into detail it will be suffi
cient to call attention to certain stones and joints. Anyone with sufficien
interest can work out the approximate order in which the stones wer
placed. Notice is directed to corner 1-2-3, base of No. 4, corner 5-6-7

239

The first page of Jessup's article on Inca masonry.

this into his first book which was published in 1955 by Citadel Press of New York City, a major publisher of books on the occult at the time. It was called *The Case for the UFO*. It came out first as a hardback book and then as a pulp paperback from Bantam Books.

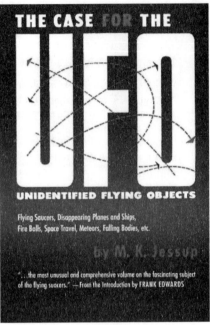

THE CASE FOR THE

UFO

UNIDENTIFIED FLYING OBJECTS

Flying Saucers, Disappearing Planes and Ships, Fire Balls, Space Travel, Meteors, Falling Bodies, etc.

by M. K. Jessup

"...the most unusual and comprehensive volume on the fascinating subject of the flying saucers." —From the Introduction by FRANK EDWARDS

The dust cover of Jessup's 1955 book.

In the book Jessup argued that UFOs represented a mysterious subject worthy of further study. Jessup speculated that UFOs were "exploratory craft of a 'solid' and 'nebulous' character." Jessup also "linked ancient monuments with prehistoric superscience" years before similar claims were made by Erich von Däniken in *Chariots of the Gods?* which first came out in German in 1968. Similar books by the French author Robert Charroux came out in the late 1960s and early 70s. In 1957 a book by George Hunt Williamson (who also used the pen name Brother Philip) called *Road in the Sky* was published that also promoted the Space Gods theory along with the megalithic ruins in Peru. Jessup's first book had come out two years earlier.

Jessup wrote three further flying saucer books, *UFOs and the Bible* and *The UFO Annual* that came out in 1956 and *The Expanding Case for the UFO* which came out in 1957. The latter suggested that transient lunar phenomena were somehow related to UFOs in the earth's skies. Jessup's main flying saucer scenario resembled that of the Shaver Mystery that was published in *Amazing Stories* magazine—namely, that "good" and "bad" groups of space aliens have been meddling with terrestrial affairs. Like most of the writers on flying saucers and the so-called contactees that emerged during the 1950s, Jessup displayed familiarity with the alternative mythology of human prehistory developed by Helena Blavatsky's books on Theosophy, which included tales of lost continents such as Atlantis, Mu, and Lemuria.

Jessup's life was suddenly to change when Carl Allen wrote to Jessup in 1955 shortly after *The Case for the UFO* had first come out. Jessup wrote a quick note in return, and continued to work on his other books for Citadel and lecture around the US on the topic of UFOs and Einstein's Unified Field Theory. Allen's letters to Jessup also referenced the Unified Field Theory which was important in the propulsion of the flying saucers.

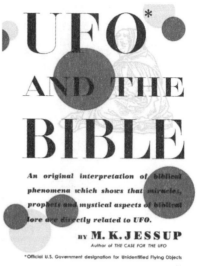

Jessup would tell audiences at the time, "If the money, thought, time, and energy now being poured uselessly into the development of rocket propulsion were invested in a basic study of gravity, beginning perhaps with continued research into Dr. Einstein's Unified Field concepts, it is altogether likely that we could have effective and economical space travel, at but a fraction of the costs were are now incurring, within the next decade."

Carl Allen attended at least one of Jessup's talks during 1955/56 and wrote Jessup a second letter. It was postmarked from Miami, on the letterhead of the Turner Hotel in Gainesville, Texas, and carried a Pennsylvania rural address for return.

Jessup tried to publish more books on the subject of UFOs, but was unsuccessful. He lost his publisher and experienced a succession of downturns in his personal life.

Morris Jessup photo on the dust jacket of *The Case for the UFO*.

The Mysterious Death of Morris Jessup

Jessup attempted to make a living writing on the subject of UFOs, but his follow-up books did not sell very well, and his publisher rejected several other manuscripts. In 1958 his wife left him, and he traveled to New York City; his friends described him as being somewhat unstable. He thought that his mail was being opened and that he might be being followed. After returning to Florida, he was involved in a serious car accident and was slow to recover, apparently increasing his despondency. This "accident" may have been on purpose and it may have been the first attempt on Jessup's life. We just do not know.

On April 20, 1959, in Dade County, Florida, Jessup's car was found along a roadside with Jessup dead inside. A hose had been wired to the exhaust pipe and run into a rear window of the vehicle, which had filled with toxic fumes when turned on. The death was ruled a suicide. Friends allegedly said that he had been extremely depressed, and had discussed suicide with them for several months, but this may just be hearsay.

Jessup's death would get rolled into more conspiracy theories surrounding the Philadelphia Experiment with some believing that "[t]he circumstances of Jessup's apparent suicide [were] mysterious." The William L. Moore and Charles Berlitz book, *The Philadelphia Experiment: Project Invisibility*, put forward the conspiracy theory that his death was connected to his knowledge of the Philadelphia Experiment. Friends interviewed in the same book thought the bizarre letters from Carl Meredith Allen may have initiated a decline in Jessup's mental state leading to his eventual suicide.

Gray Barker reports in his 1963 mimeographed book *The Strange Case of Dr. M.K. Jessup* that Jessup first learned of the annotated copy when he was talking to Mrs. Walton Colcord John, director of the *Little Listening Post,* a UFO and New Age publication in Washington DC sometime in the early 1960s. Speaking over the telephone, Mrs. John told author Brad Steiger of a strange rumor going around, to the effect that somebody had sent a marked-up copy to Washington, and that the government had gone to the expense of mimeographing the entire book, so that all the underlinings and notations could be added to the original text. This was being circulated rather widely, she told him, through military channels.

In late October of 1958 Jessup travelled from Indiana to New York, and sometime around Halloween, Jessup contacted the British Zoologist Ivan T. Sanderson, the founder of the Society for the Investigation of the Unexplained (SITU) at his home and offices in New Jersey. During this meeting, Jessup told Sanderson his bizarre story, and gave a copy of the Varo Edition to him, one of three given to him by ONR.

Jessup also confided in Sanderson things of a confidential nature, probably about the Philadelphia Experiment and time travel. Jessup also apparently thought he was being followed and his mail was being tampered with.

Jessup had apparently made several further trips to the offices of ONR after the printing of the Varo Edition. Hoover and Shelby of the ONR reportedly made efforts to track down Allende/Allen, with Hoover checking out the rural Pennsylvania address, which turned out to be bogus.

Meanwhile, Jessup had disappeared, and his publisher made

DADE COUNTY

DEPARTMENT OF PUBLIC HEALTH

1350 N W FOURTEENTH STREET

MIAMI 35, FLORIDA

T E CATO M D M P H
DIRECTOR

STATE BOARD OF HEALTH
BUREAU OF VITAL STATISTICS

CERTIFICATE OF DEATH

FLORIDA

STATE FILE NO.

REGISTRAR'S NO. **2569**

PLACE OF DEATH COUNTY: Dade	CODE NO: 23 X X	USUAL RESIDENCE STATE: Fla. COUNTY: Dade	
CITY TOWN OR LOCATION: Rural	IS PLACE OF DEATH INSIDE CITY LIMITS? YES [] NO [X]	CITY TOWN OR LOCATION: Coral Gables	RESIDENCE INSIDE CITY LIMITS? YES [X] NO [] ON A FARM? YES [] NO [X]
NAME OF HOSPITAL OR INSTITUTION: Dade County Park	LENGTH OF STAY IN TA: Minutes	STREET ADDRESS: 1531 Saragossa	

NAME OF DECEASED (Type or print): MORRIS K. JESSUP — DATE OF DEATH: April 20, 1959

SEX: Male	COLOR OR RACE: White	MARRIED [] NEVER MARRIED [] WIDOWED [] DIVORCED []	DATE OF BIRTH: March 2, 1900	AGE: 59

USUAL OCCUPATION: Professor of astronomy — KIND OF BUSINESS OR INDUSTRY: University — BIRTHPLACE: Indiana — CITIZEN OF WHAT COUNTRY: U.S.A.

FATHER'S NAME: Unobt. — MOTHER'S MAIDEN NAME: Unobt.

WAS DECEASED EVER IN U.S. ARMED FORCES? Yes — SOCIAL SECURITY NO.: WWI — INFORMANT'S SIGNATURE: ... Address: 1531 Saragossa...

CAUSE OF DEATH — IMMEDIATE CAUSE: Acute carbon monoxide intoxication

PART II OTHER SIGNIFICANT CONDITIONS CONTRIBUTING TO DEATH BUT NOT RELATED TO THE TERMINAL DISEASE CONDITION GIVEN IN PART I(a)

WAS AUTOPSY PERFORMED? YES [] NO [X]

ACCIDENT [] SUICIDE [X] HOMICIDE []	DESCRIBE HOW INJURY OCCURRED: Deceased inhaled auto exhaust.	
TIME OF INJURY: p.m. p = 4/20/59	PLACE OF INJURY: Station wagon -County Park	CITY TOWN OR LOCATION: Rural COUNTY: Dade STATE: Florida

WHILE AT WORK [] NOT WHILE AT WORK []

973

I HEREBY CERTIFY ... Death occurred at 6:30 p.m. ... on the date stated above, and to the best of my knowledge, from the causes stated

SIGNATURE: ... — MEDICAL EXAMINER'S OFFICE — DATE SIGNED: 4/21/59

BURIAL CREMATION REMOVAL ... Apr. 22, 1959 U. of Miami Medical School Coral Gables, Fla.

FUNERAL DIRECTOR: Olin E. Alleguez ... DATE REC'D BY LOCAL REG: APR 22 1959 ... Ethel Henshaw

THIS IS A TRUE ... COPY OF THE LOCAL REGISTRARS RECORD OF DEATH.

SEAL

This record VOID unless the seal of the Deputy-Registrar appears thereon.

Ethel Henshaw
DIRECTOR AND DEPUTY-REGISTRAR DIST. #23
BUREAU OF VITAL STATISTICS
DADE COUNTY HEALTH DEPARTMENT
MIAMI, FLORIDA

Morris Jessup's death certificate.

efforts to contact him. Eventually, it was discovered that Jessup had driven directly to Florida from New York, where he had intended on living. He was apparently fleeing his home in Indiana, possibly believing he was being watched. He had had a car accident as well, but had survived. He had been in the hospital for an extended period.

Then, on April 20, 1959, he was found dead in his parked car in the rural Dade County Park near his Coral Gables home. A rubber hose ran from his exhaust pipe into the nearly closed back window of the car. His death was ruled as a self-inflicted carbon-monoxide poisoning. More on that in the next chapter.

But some, including Ivan T. Sanderson, believed that Jessup had not killed himself, but instead had been "suicided" by the Men in Black who had taken a deep interest in his contact with Carlos Allende/Carl Allen and his knowledge of the so-called Philadelphia Experiment.

Indeed, why would Jessup take the time to meticulously wire the hose to the tailpipe and put it into the back window? Most suicides in this manner do not go to the extra trouble to get wire to use in securing the hose to the tailpipe.

In fact, if his publisher was trying to contact him it was probably to approve a new book. Citadel Press continued to publish UFO and occult books for decades and it seems that they had a good relationship with Jessup, rather than a negative one.

Bill Moore in *The Philadelphia Experiment* says that it was reported that Jessup had a massive amount of alcohol in his system, so much so he would not have been able to wire the hose to the tailpipe.

Says Moore on page 151 of *The Philadelphia Experiment*[7]:

> The first bit of evidence comes by way of Mrs. Anna Genslinger of Miami, who, along with a police lieutenant friend of hers, has obtained access to the Dade County, Florida, medical examiner's files on the Jessup case. It was in these records that she discovered that at the time of his death Jessup's blood was virtually saturated with what would normally be considered more than a lethal amount of alcohol. According to Mrs. Genslinger, Jessup was also

taking a medication at the time which, when combined with such an amount of alcohol, could have been immediately fatal or at least would have been far more than enough to incapacitate him totally. He would have been completely unable even to get in a car under his own power, much less drive one several miles to a county park, write a suicide note, and then attach a hose to the exhaust pipe of the vehicle. Furthermore, no autopsy was ever performed on the body—an unusual occurrence in a suicide case. Of course, none of this constitutes conclusive evidence for murder, but it is nonetheless of interest.

Equally interesting is a comment made to the author by James R. Wolfe, a freelance writer and researcher who had himself spent some time looking into the Allende mystery. Wolfe had started to write a book on the topic when he became a mystery himself by turning up missing before it was completed. Even stranger is the fact that it was Wolfe's disappearance which led to the discovery of the additional evidence provided in the next chapter. Before his disappearance, however, Wolfe had exchanged with the authors useful bits and pieces of information. In discussing Jessup's possible murder, Wolfe, a former Navy man, indicated that he didn't believe it at first but later became convinced. He went on to say that, "the big reason for the continued secret classification on the Philadelphia Experiment is not the damage that knowledge of it would do to the Navy—but the damage it would do to the image of *an individual.*" That individual, according to Wolfe, carried more than enough clout not only to arrange for Jessup's murder, but to see that the job was neatly done. He did not, however, identify the individual he had in mind.

Looking in other directions, perhaps yet another hint as to the reason for the continued secrecy has already been provided to us by one of Charles Berlitz's informants. This informant, who emphatically declined to be named, confided to Berlitz that he had seen highly classified documents in the Navy files in Washington, D.C., which indicated that at least some phases of the experiments are *still* in progress.

In addition, scientific units in private universities, some possibly funded by the government, are reported to be pursuing research in magnetic teleportation, with the attendant invisibility as part of the experiment. Some recent reports place such experimentation as having taken place at Stanford University Research Facility at Menlo Park, Palo Alto, California, and at M.I.T. in Boston. However, in the words of one informant—M. Akers, a psychologist in San Jose, California—such magnetic experiments "are frowned upon because they have detrimental effects on the researchers conducting the experiments."[7]

Moore concludes this chapter by musing whether the Philadelphia Experiment was still happening. His discussion of Wolfe is very interesting and Moore continues the discussion in the next and final chapter to the book.

Moore believes that Jessup was murdered. If so, by whom? By the Men in Black? By agents of Naval Intelligence? By the FBI? Was it possibly by the Defense Industrial Security Command (DISC) to be discussed in a future chapter? Perhaps these agents of DISC were essentially the "Men in Black." And, if this is the case, the mysterious individual that Wolfe would not name might be Wernher von Braun.

Moore says his last contact with Wolfe was in 1977 and in February 1978 he began hearing reports that he had disappeared. One of the reports came from a business associate in the publishing world. Moore eventually sent a letter to Wolfe which came back after a few days later marked "Not at this address."

Moore then got a phone call in early May 1978 from Michelle Alberti who identified herself as secretary of CUFORN, Inc., a Canadian psychic research group in Willowdale, Ontario. She

> **SAUCER MAN SUICIDE**
>
> MIAMI, April 21 (UP) — Police said tonight that Morris K. Jessup, 59, widely known writer on flying saucers, took his own life with a hose linked to his automobile's exhaust pipe. Jessup, a mathematician and astronomer, was a former University of Michigan and Drake University instructor.

A short newsclipping of Jessup's death.

FAMOUS IOWANS

UFO researcher Jessup remains a big mystery

By TOM LONGDEN
REGISTER STAFF WRITER

The M.K. Jessup story, including his mysterious death, is absolutely spooky.

The man who once lived in Iowa is an icon of the unexplained, a major figure in paranormal fields, especially unidentified flying objects, or UFOs. He even coined the word UFOlogy (u-fol-ogy).

Jessup, who taught in Des Moines, is forever linked to the Philadelphia Experiment, a bizarre incident that has never been fully explained despite countless articles and books.

The event allegedly occurred in October 1943, when a Navy ship, the USS Eldridge, was the subject of a top-secret war experiment. Through electromagnetic or similar technology, such as a variation of Einstein's Unified Field Theory, the ship became invisible and then was supposedly teleported from a dock in Philadelphia, Pa., to one in Norfolk, Va., and then back again in a matter of minutes.

The episode is recounted in such books as "The Philadelphia Experiment: Project Invisibility" (1979), written by William L. Moore in collaboration with Charles Berlitz.

The Navy, which enlisted Jessup's help during an investigation after receiving a mysterious annotated copy of Jessup's first book, says such an experiment never happened.

Morris Ketchum Jessup was born March 20, 1900, in Rockville, Ind., but little is known about his early life.

Upon graduating from high school, Jessup immediately went into the Navy, serving in 1918-19, and then enrolling at the University of Michigan at Ann Arbor, also in 1919.

With time out for travel, he graduated with a bachelor's degree in science in 1926 and earned a master's degree the next year. He studied for a doctorate degree but did not graduate.

Jessup became fascinated with astronomy and astrophysics, and a circuitous route brought him to Des Moines. In 1932, he became a professor at Drake University.

Today, the name Morris Jessup does not ring a bell with many people at Drake, but Mark Stumme of Drake's Cowles Library has dug up some information on the scientist's years in Des Moines.

An article Oct. 6, 1932, in the Drake

SPECIAL TO THE REGISTER

Morris K. Jessup
Astronomer, mathematician, UFO researcher and lecturer

Photos of Jessup are rare. Here Jessup is shown in a picture that appeared in the Drake University yearbook. A pioneer in the study of UFOs, he always called for further research, saying "there is much to learn." Many of Jessup's theories predated those of Erich Von Daniken, who achieved popularity in 1968 for his book "Chariots of the Gods?"

The Jesup angle
FAMOUS UNCLE: Morris K. Jessup was named for his uncle, Morris K. Jesup (spelled with just one s). Jesup, a self-made man, railroad baron, banker and philanthropist, is honored with a cape in Greenland named for him. Jesup had financed several polar expeditions.

ALSO AN IOWA TOWN: The town of Jesup is also named for Jesup because he was part owner of a rail line running from Dubuque to Jesup, just east of Waterloo.

More Famous Iowans on the Web
Go to **DesMoinesRegister.com/Iowans** to find past articles about famous Iowans.

Do you have suggestions?
Send your suggestions for other famous Iowans to feature to tlongden@dmreg.com.

Times-Delphic, the campus newspaper, said "Morris K. Jessup comes to Drake as instructor in mathematics and astronomy with a colorful background and a wide and varied experience. While astronomy is his chief interest, amateur photography, radio and aviation are his hobbies."

An article two weeks later said he planned to be at the university's observatory on public nights in order to make astronomy popular with the public.

A Des Moines Register article May 16, 1933, said Jessup would lead 15 students on a study tour of six Latin American nations that summer, and said he had previously made three trips to that area.

"He was an engineer in a Carnegie Institution expedition, later an expert in a crude rubber survey for the U.S. Department of Agriculture, and last summer he and Mrs. Jessup traveled extensively in South America en route to South Africa," the article said.

Jessup was pictured in the 1935 Drake yearbook, the Quax, with other members

of the Drake math club.

In the 1940s and 1950s, Jessup, who had left Iowa, took a particular interest in the new field of UFOs — flying saucers. With a move to Washington, D.C., in the 1950s, Jessup began writing his first book in 1954. "The Case for the UFO" was published in 1955 and had moderate success, followed by "UFO and the Bible" (1956) and "The UFO Annual" (1956).

His last book was "The Expanding Case for the UFO" (1957).

On the evening of April 20, 1959, Jessup was found moments from death, slumped over the wheel of his station wagon in a Dade County park near his Florida home. A hose attached to the car's exhaust pipe had filled the car's interior with carbon monoxide. No suicide note was ever found, and no autopsy was performed.

Jessup reportedly was working on a new book about the Philadelphia Experiment. Was Jessup silenced — murdered because he knew too much? The controversy continues.

Morris Jessup's page on the Famous Iowans website.

explained that during her group's investigation of the Philadelphia Experiment she had learned about James Wolfe who was said to have information on the matter. While trying to locate Wolfe, she was dismayed to discover that he had disappeared. She then said that further inquiries indicated that he might be dead.

Immediately suspecting another "Jessup-type incident" she called Gray Barker in Clarksburg, West Virginia, to see what he knew. Apparently Barker did not know anything and gave her the phone number of Moore, and so they were now talking. Alberti told Moore a strange story about a man seeing a flying saucer near her house and then encountering a four-foot-tall being in a spacesuit and helmet standing near a craft. Later the person claimed that top brass from "the Canadian Forces in Ottawa, the United States Airforce, Pentagon, and from the Office of Naval Intelligence" came to the home and interviewed them and then answered any questions the person may have had.

Moore finds this quite interesting as it demonstrates a keen knowledge and curiosity in regard to UFOs on the part of the US Navy, Pentagon and Air Force, not to mention the Canadian military.

Moore says that the mystery of James Wolfe continues. He was still missing as Moore finished the book in 1979. He was presumed dead by most people. We seem to have another victim of the Philadelphia Experiment. An Internet search for him does not bring up any person identified as him, nor was his book ever published. He seems to have genuinely vanished. Was he getting too close to the truth and becoming a nuisance to certain powers and individuals like Naval Intelligence and Wernher von Braun?

It is interesting to speculate if Carl Allen/Carlos Allende was also a target of these supposed assassins? Indeed, it seems likely that he had been but his vagabonding and gypsy ways made him a very difficult person to find. When people did meet him, such as Jim Lorenzen at the APRO headquarters in Tucson, Allen/Allende would just suddenly show up unannounced.

Jessup was "murdered" in 1959 and Wolfe disappeared in 1978. But between those dates there may have been another murder because of the Philadelphia Experiment. That murder may have been the murder of Ian Fleming in 1964.

Ian Fleming in the early 1960s.

Ivan T. Sanderson, Ian Fleming, and Aleister Crowley

In a curious sidenote, the popular James Bond "007" author, Ian Fleming, is strangely linked to the Philadelphia Experiment. According to Preston B. Nichols and Peter Moon in their book *Pyramids of Montauk*[33] Ian Fleming knew certain information about the Rainbow Project, the secret endeavor that was to ultimately lead into what is known today as the Philadelphia Experiment.

Fleming had worked with Aleister Crowley on the Rainbow Project (acknowledged by the Navy to be a plan to defeat the Axis in WWII), his part being a secret mission to meet with Karl Haushofer of the Nazi party in order to get him to convince Rudolf Hess to defect. Fleming met with Haushofer in Lisbon, Portugal, early in the war and persuaded the influential German occultist to talk with Hess on behalf of Crowley. Both Haushofer and Hess admired Aleister Crowley a great deal, according to Nichols and Moon.

In August of 1964, say Nichols and Moon, Fleming was planning to fly from his home in Jamaica to New Jersey to meet with Ivan T. Sanderson, a biologist and former British Intelligence agent. As a zoologist, Sanderson had written a

Ian Fleming as a Naval Commander.

82

Ivan T. Sanderson in the mid-1960s.

number of books, including one on bigfoot and yeti, and appeared frequently on radio shows and Johnny Carson's *Tonight Show.*

Sanderson had become intensely interested in the paranormal, and was a personal friend of Morris K. Jessup, as noted above. Sanderson often talked with his friends about UFOs, the Philadelphia Experiment, cryptozoology and other arcane subjects. Like Fleming, Sanderson was a British expatriate.

According to Nichols and Moon, Sanderson had been corresponding with Fleming, exchanging key information about the Rainbow Project, perhaps the involvement of Crowley, and how it related to the Philadelphia Experiment. Perhaps Fleming had some inside information as to the secret technology to make battleships invisible, teleport them, and ultimately on UFO technology in general.

In any event, Fleming never made it to his rendezvous with Sanderson. He died of a sudden heart attack on August 12, 1964 at his home in Jamaica. Nichols and Moon point out that this is

Aleister Crowley.

the 21st anniversary of the Philadelphia Experiment (August 12, 1943, however other dates have been given). Was Ian Fleming killed because he knew too much about such things as the Kennedy assassination, the Philadelphia Experiment and the genesis of UFO technology?

Was there More than One Morris K. Jessup?

Lastly, we have to broach the subject of "Was there more than one Morris K. Jessup?" Indeed there is the possibility that the real Morris K. Jessup did not die in 1959. Why do I say this?

84

For one thing there are very few photos of Morris Jessup. However, there are about five that we know about. The first photo we have of him is as a seaman in the US Navy. It may have been taken around 1919. Then we have the photo of Jessup with Carl LaRue and James Weir in Brazil in 1923.

After that we have a photo from about 1955 that was used as a jacket photo on the hardback edition of *The Case for the UFO*. As in the other two photos of Jessup, he is heavyset and has dark circles around his eyes. However, there is one

Said to be a photo of Jessup.

other photo of Morris K. Jessup and this is of a man who wears glasses and seems to have a different hairline from the man in the other photos, and does not have the dark circles around the eyes.

Is this photo really a photo of Morris K. Jessup? It looks like a photo of someone else, not of the Morris Jessup who went to Brazil and wrote books. Were there two Morris Jessup's? If so, why would the other one be?

Why are there so few photos of Morris Jessup? He was a photographer himself. Was he always behind the camera? Was there an effort on his part to not have his photograph taken? This is common among intelligence agents and other operatives. Was he an intelligence operative of some kind? What of the second person claiming to be Jessup? Was he also an operative? As an example, there were known to be several different Lee Harvey Oswalds running around in Dallas and Mexico City just prior to the JFK assassination.

So, by extension, we have the possibility that the real Morris K. Jessup was not "suicided" in 1959 and that the body in the car was someone else. No one who knew Jessup identified his body and no autopsy was performed. It seems like it was treated as an open and shut case, where a body was found and a death certificate was given without any other investigation.

This is all pure speculation, and from a conspiracy angle we must honestly fall back on the mainstream story of Jessup being extremely incapacitated and incapable of connecting a hose to a tailpipe.

Just as startling is the revelation that William Moore made in 1985 in his catalog of copied documents that he sold for $10 to $20 apiece. The revelation that Moore made was that he had obtained documents through the Freedom of Information Act that said that "Jessup had made death threats against the President of the United States." A copy of this document was for sale. I ordered one from William Moore.

As one might imagine, this is a very serious accusation and one that seems fantastically bizarre. No one would think that Jessup would make death threats against the president, who was Dwight Eisenhower at the time. It seems utterly ridiculous that there would be an FBI file or such that claimed that Jessup was making or had made death threats against the president.

The thinking here is that the worst thing that can be in an official police or FBI document is that a person made a death threat against the president. Such a claim would literally turn every officer of the law against you and a manhunt of huge proportion would immediately follow. It is the worst form of "official slander" that anyone could make up about you. You are literally likely to be gunned down by the police, FBI or Secret Service if this false claim is made about you. It is perhaps the highest crime you could be accused of in the United States.

So, why was it claimed that Jessup had made the obviously false move of declaring, somehow, that he wanted to kill the President of the United States? It would seem to tarnish the reputation of Jessup in the worst possible way to people who did not know him. Police officers, detectives, FBI agents and such who viewed such a document would assume it was true and this would not reflect well on Morris Jessup, whether he was alive or dead. It seems to be one more clue that someone or some organization, high up in the government chain of command, had wanted to tarnish Jessup's image and have him killed. Was this person Wernher von Braun?

Chapter 5

What Was Project Rainbow?

Everyone was hangin' out
Hangin' up and hangin' down
Hangin' in and holding fast
Hope our little world will last
—*Ship of Fools*, The Doors

According to Carl Allen the project that turned the USS *Eldridge* invisible was called Project Rainbow. The popular term has become the Philadelphia Experiment mainly because of the movie—the story was adapted into a 1984 time travel film called *The Philadelphia Experiment*, directed by Stewart Raffill. Though only loosely based on the prior accounts of the "Experiment," it served to dramatize the core elements of the original story.

In 1965 Vincent Gaddis published the book *Invisible Horizons: True Mysteries of the Sea*. In it he recounted the story of the experiment from the Varo annotations. George E. Simpson and Neal R. Burger published a 1978 novel titled *Thin Air*. In this book, set in the present day, a Naval Investigative Service officer investigates several threads linking wartime invisibility experiments to a conspiracy involving matter transmission technology.

Large-scale popularization of the story came about in 1979 when the author Charles Berlitz, who had written a best selling book on the Bermuda Triangle, and his co-author, ufologist William L. Moore, published *The Philadelphia Experiment: Project Invisibility*. The book expanded on the bizarre happenings in 1943, lost unified field theories by Albert Einstein, government cover-ups, and the mystery of Jessup's death.

87

Moore and Berlitz devoted one of the last chapters in *The Philadelphia Experiment: Project Invisibility* to "The Force Fields Of Townsend Brown," namely the experimenter and then-U.S. Navy technician Thomas Townsend Brown. Paul LaViolette's 2008 book *Secrets of Antigravity Propulsion* also recounts the mysterious involvement of Townsend Brown in Navy experiments.

But was there a Project Rainbow during WWII? It has been said that a Project Rainbow was essentially the overall plan to defeat the Nazis during WWII. But the Navy and the National Archives has said that there was no Project Rainbow during WWII.

When William Moore and Charles Berlitz contacted the National Archives in 1977 they were told that there was no record of a Project Rainbow and no knowledge of an index of projects. Then when the name "Rainbow" was given to them the staff produced a 1941 Inter-Services Code-Word Index. This index had about 10,000 code words in it and "Rainbow" was indeed in the index. It was entry 7470.

So, this was proof to Moore and Berlitz that Rainbow had indeed been a code word during WWII. Did it refer to the Philadelphia Experiment? That is what Carl Allen had said.

Secret Projects of the US Navy

It is known that the Germans had secret experiments and created experimental aircraft during the war, including rockets, jet fighters, super long distance aircraft and even flying saucers. Like the US military, the German military was broken into competing factions such as the Navy, Army, and the Air Force. Add to that the SS battalions that were completely separate from the ordinary military.

Similarly, the US Navy has had many secret projects over the years. During the important years around WWII the Navy had a number of secret projects. Some of the experiments had to do with the degaussing of ships. Some experiments may have evolved into efforts to turn a ship invisible. It only seems natural the Navy would want such a thing and would do experiments in this direction. Such experiments probably involved wrapping copper cables around a ship, in a similar manner that copper cables are wound around an electric motor.

MOST SECRET

Copy No.

**Attention is drawn to the Penalties attaching
to any infraction of the Official Secrets Acts**

INTER-SERVICES
CODE-WORD INDEX

This Index will be kept
in a safe when not in use

Issued under the authority of
the Inter-Services Security
Board, War Office.

1st September, 1941

C.C.S. 368/15.

The Inter-Services Code-Word Index for 1941. The code word Rainbow appears on
page 76 demonstrating that Rainbow was part of a system of projects and operations.

As related earlier, the experiment was allegedly based on an aspect of some unified field theory, a term coined by Albert Einstein to describe a class of potential theories; such theories would aim to describe—mathematically and physically—the interrelated nature of the forces of electromagnetism and gravity. Einstein believed that magnetism, electricity and gravity all manifested from one source: the unified field. Therefore, just like there is an equation for electromagnetism, there should be an equation for electrogravity and an equation for magnetogravity.

We will discuss degaussing shortly, but let us first look at some of the other operations throughout the years that have been called Project Rainbow.

Some of the more recent Project Rainbows are essentially programs to help house LGBTQ homeless people. A Chicago Cook County operation called Project Rainbow was to provide literacy videos to 3- to 6-years olds.

Project RAINBOW was the name given by the CIA to a research project in 1956 aimed at reducing the radar cross section of the Lockheed U-2 high altitude spy plane and lowering the chance that it would be detected and tracked by Soviet radars during overflights of the USSR. However, the Soviets were able to continue tracking the U-2 flights in spite of experimentation with various technological fixes.

Wikipedia says that the U-2 was developed by Lockheed Aircraft Corporation for the CIA to perform aerial reconnaissance over flights of the Soviet Union. Project director Richard M. Bissell assured President Dwight Eisenhower that the aircraft's high altitude (70,000 feet or 21,000 meters) would render it invisible to Soviet radars. However, the earliest flights in July 1956 were, in fact, tracked by the Soviets. On July 5, 1956, an A-100 "Kama" radar detected Carmine Vito as he flew over Smolensk in the U-2, en route to Moscow. The operators even calculated his altitude as twenty kilometers (66,000 feet), which was later rejected by experts because they did not believe that an aircraft could fly that high.

In mid-August 1956, Bissell assembled a group of advisers to begin work on solving the radar-tracking problem. Among the group were Edwin H. Land, founder of the Polaroid Corporation

Ciphering Numeral.	Code Word.	Block Number.	Ciphering Numeral.	Code Word.	Block Number.
7400	PUSSYFOOT	374	7450	RACQUET	266
7401	PUTNEY	590	7451	RADCLIFFE	748
7402	PUTRID	287	7452	RADFORD	854
7403	PUTTENHAM	804	7453	RADIAL	266
7404	PUZZLE	181	7454	RADIATION	324
7405	PYGMALION	164	7455	RADIATOR	491
7406	PYRAMID	170	7456	RADIO	2
7407	PYRITES	979	7457	RADISHES	304
7408	PYRRHUS	170	7458	RADIUM	430
7409	PYTHAGORAS	594	7459	RADNAGE	722
7410	PYTHON	288	7460	RADNOR	883
7411	QUADRANGLE	149	7461	RAEBURN	748
7412	QUADRUPED	548	7462	RAFFIA	266
7413	QUAGGA	619	7463	RAFFLE	422
7414	QUAGMIRE	126	7464	RAFTER	287
7415	QUAINTON	871	7465	RAGAMUFFIN	126
7416	QUARRY	143	7466	RAGMAN	181
7417	QUARTER	454	7467	RAGOUT	432
7418	QUARTERMAIN	388	7468	RAILWAY	33
7419	QUEEN	17	7469	RAIMENT	933
7420	QUEENBEE	643	7470	RAINBOW	334
7421	QUENCH	234	7471	RAINHAM	827
7422	QUICK	665	7472	RAISIN	229
7423	QUICKFIRE	104	7473	RAKEOFF	988
7424	QUICKLIME	548	7474	RAKISH	544
7425	QUICKSAND	561	7475	RALLY	188
7426	QUINCES	304	7476	RALPH	44
7427	QUINN	365	7477	RALSTON	827
7428	QUINTAL	123	7478	RAMBLER	431
7429	QUINTIN	266	7479	RAMESES	600
7430	QUIRK	585	7480	RAMIFICATION	63
7431	QUISLING	119	7481	RAMMER	288
7432	QUITMAN	659	7482	RAMONA	457
7433	QUIVER	129	7483	RAMOSE	158
7434	QUIXOTE	29	7484	RAMPAGE	434
7435	QUIZZY	173	7485	RAMPART	556
7436	QUOIT	393	7486	RAMPION	22
7437	QUORUM	563	7487	RAMPIRE	233
7438	QUOTATION	563	7488	RAMROD	143
7439	RABBITS	323	7489	RAMSDEN	853
7440	RABBLE	123	7490	RAMSHACKLE	561
7441	RABELAIS	450	7491	RAMSHORN	854
7442	RABID	652	7492	RAMSON	267
7443	RACEFIELD	559	7493	RANCHER	534
7444	RACHEL	447	7494	RANCID	141
7445	RACING	308	7495	RANDALSTOWN	738
7446	RACKETEER	487	7496	RANDOM	451
7447	RACKHAM	883	7497	RANELAGH	590
7448	RACONTEUR	266	7498	RANGE	19
7449	RACOON	266	7499	RANJI	601

The Inter-Services Code-Word Index 1941 page 76 demonstrating that Rainbow was part of a system of projects and operations as can be seen in the second column.

and head of the Technological Capabilities Panel; Edward Purcell, a Nobel laureate physicist from Harvard; and Clarence L. "Kelly" Johnson, head of Lockheed Advanced Development Projects (ADP)—the Skunk Works.

Wikipedia says the group conducted initial discussions and then Edwin Land went to the MIT Lincoln Laboratory to recruit radar specialists for the work. The leader of the Lincoln Lab team was Franklin Rodgers, associate head of the radar division. Working in isolation from the rest of the lab, his group began trying to find ways to reduce the U-2's radar cross section. As their work progressed, they traveled to California to work with Lockheed and to various military bases to perform radar measurements of U-2s in flight.

By the fall of 1957, only months after the first deployment of the U-2, it had become obvious to Bissell and the scientific team that the treatments would only have a marginal effect on tracking, and that a new aircraft would be needed. By having anti-radar features designed into the craft from the beginning, it was hoped that the succeeding aircraft would escape detection.

By this time a large number of people had become aware of Project RAINBOW. To reduce the spread of information about the follow-up work, it was moved into a new project, GUSTO. With Project GUSTO only those with a need to know were cleared into it. The end result of GUSTO would be the Lockheed A-12 OXCART built at the Skunk Works.

So, it seems that the 1956 Project RAINBOW was a project on radar invisibility. However, it would seem that there was an earlier Project Rainbow run by the Navy during WWII doing something similar.

Has the Unified Field Theory Been Surpressed?

Carl Allen in his letters to Jessup and in notes of the Varo Edition of *The Case for the UFO* continually mentioned Einstein's unified field theory and how it was part of the Rainbow Project, which involved wrapping cables around a ship and turning on an electromagnetic field. Gravity would be involved in this as well as that is what the unified field is: a relationship between electricity, magnetism and gravity. Four-dimensional spacetime is

The Philadelphia Naval Yard during WWII.

also involved in this relationship. To revisit Wikipedia's article about the unified field theory:

By 1905, Albert Einstein had used the constancy of the speed of light in Maxwell's theory to unify our notions of space and time into an entity we now call spacetime and in 1915 he expanded this theory of special relativity to a description of gravity, general relativity, using a field to describe the curving geometry of four-dimensional spacetime.

93

The article then concludes:

Theoretical physicists have not yet formulated a widely accepted, consistent theory that combines general relativity and quantum mechanics to form a theory of everything. Trying to combine the graviton with the strong and electroweak interactions leads to fundamental difficulties and the resulting theory is not renormalizable. The incompatibility of the two theories remains an outstanding problem in the field of physics.

As we have seen, according to some accounts, unspecified "researchers" thought that some version of this unified field would enable using large electrical generators to bend light around an object via refraction, so that the object became completely invisible. The Navy regarded this as being of military value and it sponsored the Philadelphia Experiment.

Carl Allen and Al Bielik have both claimed that the USS

The USS *Eldridge* at sea in 1943.

Eldridge was fitted with the required equipment at the Philadelphia Naval Shipyard. Testing began in the summer of 1943, and it was supposedly successful to a limited extent.

No matter how fantastic Project Rainbow may seem, there seems to be no doubt that US Navy was doing research of this kind during and after WWII. We know for a fact the US Navy was winding ships with copper cables and attempting to degauss them against magnetic mines. But had the US Navy made the obvious leap in turning electromagnetic fields around a battleship into an experiment in invisibility and possibly teleportation? Why not? It seems like the obvious next step to take with this technology.

When we think about the Nazi experiments with "the Bell," and their work on the Vril, Haunebu, and Andromeda craft, and other exotic weapons—including the atomic bomb—we have to assume that the Americans, along with their allies in Britain and Canada, were working on their own advanced projects.

Indeed, with all the stories of the secret projects at Area 51 in Nevada one has to wonder how far back these secret projects go? All the way to World War II? Yes, this is obviously the correct answer.

Even a number of "anonymous" military scientists made some interesting comments to Moore and Berlitz that they included in their book *The Philadelphia Experiment*. Moore says that:

> The thrust of [Einstein's attempt to produce a] Unified Field Theory was a string of sixteen incredibly complex quantities (represented by an advanced type of mathematical shorthand known as tensor equations), ten combinations of which represented gravitation and the remaining six electromagnetism.
>
> One thing that does emerge, interestingly, is the concept that a pure gravitational field can exist without an electromagnetic field, but a pure electromagnetic field cannot exist without an accompanying gravitational field.[7]

Moore then says that an anonymous military scientist told him the following:

I think I heard they did some testing both along the river [the Delaware] and off the coast, especially with regard to the effects of a strong magnetic force field on radar detection apparatus. I can't tell you much else about it or about what the results were because I don't know. My guess, and I emphasize GUESS, would be that every kind of receiving equipment possible was put aboard other vessels and along the shoreline to check on what would happen on the 'other side' when both radio and low- and high-frequency radar were projected through the field. Undoubtedly observations would have also been made as to any effects that field might have had on light in the visual range. In any event, I do know that there was a great deal of work being done on total absorption as well as refraction, and this would certainly seem to tie in with such an experiment as this. [7]

William Moore quotes from an anonymous elderly scientist in the chapter called "The Unexpected Key." The man says:

...The idea of producing the necessary electromagnetic field for experimental purposes by means of the principles of resonance was... initially suggested by [physicist] Kent... I recall some computations about this in relation to a model experiment [i.e., an experiment conducted using scale models rather than real ships] which was in view at the time... It also seems likely to me that 'foiling radar' was discussed at some later point in relation to this project...

The initial idea seems to have been aimed at using strong electromagnetic fields to deflect incoming projectiles, especially torpedoes, away from a ship by means of creating an intense electromagnetic field around that ship. This was later extended to include a study of the idea of producing optical invisibility by means of a similar field in the air rather than in the water...

He had on one part of a sheet a radiation-wave equation, and on the left side were a series of half-finished scratches.

With these he pushed over a rather detailed report on naval degaussing equipment and poked fingers at it here and there while I marked with pencil where he pointed. Then Albrecht said could I see what would be needed to get a bending of light by, oh, I think 10 percent, and would I try to complete this enough to make a small table or two concerning it…

"I think that the conversation at this point had turned to the principles of resonance and how the intense fields which would be required for such an experiment might be achieved using this principle…

Somehow, I managed to finish a couple of small tables and a few sentences of explanation and brought all back as a memo. We went in to Albrecht, who looked it all over and said, "You did all this regarding intensities [of the field] at differing distances from the [ship's] beam, but you don't seem to pick up anything fore and aft." … All I had was the points of greatest curvature right off the ship's beam opposite this equipment…

What Albrecht wanted to do was to find out enough to verify the strength of the field and the practical probability of bending light sufficiently to get the desired 'mirage' effect. God knows they had no idea what the final results would be. If they had, it would have ended there. But, of course, they didn't.

I think the prime movers at this point were the NDRC and someone like Ladenburg or von Neumann who came up with ideas and had no hesitancy in talking about them before doing any computations at all. They talked with Einstein about this and Einstein considered it and took it far enough to figure out the order of magnitude he would need on intensity, and then spoke to von Neumann about what would be the best outfit to look into it as a practical possibility. That's how we got involved in it…

I can also remember a point a little later when I suggested in a meeting of some sort that an easier way to make a ship vanish was a light air blanket, and I wondered why such a fairly complicated theoretical affair was

Einstein at a meeting with US Naval Officers in 1943.

under consideration. Albrecht took off his glasses at that point and commented that the trouble with having me at a conference was that I was good at getting them off the topic...

I do remember being in at least one other conference where this matter was a topic on the agenda. During this one we were trying to bring out some of the more obvious—to us—side effects that would be created by such an experiment. Among these would be a 'boiling' of the water, ionization of the surrounding air, and even a "Zeemanizing" of the atoms; all of which would tend to create extremely unsettled conditions. No one at this point had ever considered the possibility of interdimensional effects or mass displacement. Scientists generally thought of such things as belonging more to science fiction than to science in the 1940's. In any event, at some point during all this, I received a strong putdown from Albrecht, who broke in with something to the effect that "Why don't you just leave these experimental people alone so they can go ahead with their project. That's what we have them for!"

One of the problems involved was that the ionization created by the field tended to cause an uneven refraction of the light. The original concepts that were brought down to us before the conference were laid out very nicely and neatly, but both Albrecht and Gleason and I warned that according to our calculations the result would not be a steady mirage effect, but rather a 'moving back and forth' displacement caused by certain inherent tendencies of the AC field which would tend to create a confused area rather than a complete absence of color. "Confused" may well have been an understatement, but it seemed appropriate at the time. Immediately out beyond this confused area ought to be a shimmering, and far outside ought to be a static field. At any rate, our warning on this, which ultimately went to NDRC, was that all this ought to be taken into account and the whole thing looked at with some care. We also felt that with proper effort some of these problems could be overcome... and that a resonant frequency could probably be found that would possibly control the visual apparent internal oscillation so that the shimmering would be at a much slower rate... I don't know how far those who were working on this aspect of the problem ever got with it...

Another thing I recall strongly is that for a few weeks after the meeting in Albrecht's office we kept getting requests for tables having to do with resonant frequencies of light in optical ranges. These were frequently without explanation attached, but it seems likely that there was some connection here...[7]

In another testimonial recounted by Moore, this time by an ex-military guard, reported to Moore by Patrick Macey, as told to him by a workmate named "Jim" in the summer of 1977:

> I was a guard for classified audiovisual material, and in late 1945 I was in a position, while on duty in Washington, to see part of a film viewed by a lot of Navy brass, pertaining to an experiment done at sea. I remember only part of the film, as my security duties did not permit me to sit and look at it like the others. I didn't know what was going on in the film, since it was without commentary. I do remember that it concerned three ships. When they rolled the film, it showed two other ships feeding some sort of energy into the central ship. I thought it was sound waves, but I didn't know, since I, naturally, wasn't in on the briefing.
>
> After a time the central ship, a destroyer, disappeared slowly into a transparent fog until all that could be seen was an imprint of that ship in the water. Then, when the field, or whatever it was, was turned off, the ship reappeared slowly out of thin fog.
>
> Apparently that was the end of the film, and I overheard some of the men in the room discussing it. Some thought that the field had been left on too long and that that had caused the problems that some of the crew members were having.[7]

Moore then mentions two scientists who were apparently part of Project Rainbow:

> The names of several scientists have come up in connection with the revival of such a project. Two government-employed scientists named Charlesworth and Carroll were reportedly responsible for installing the auxiliary equipment on the DE 173 [USS *Eldridge*] and participated in the experiment, noting the neuronal damage "due to diatheric" effect because of the magnetic oscillation of the magnetic field. [7]

100

And finally Moore talks about a Navy officer named Victor Silverman who apparently took part in some sort of experiment:

> Victor Silverman, now living in Pennsylvania and still mindful of wartime security regulations and afraid of possible consequences, got in touch with the authors [Berlitz & Moore] through a third party when he first heard about the publication of a book about the DE 173. He speaks from personal experience: "I was on that ship at the time of the experiment."
>
> At the outbreak of WWII Silverman enlisted in the Navy. He, along with about 40 others, was destined to become part of a special secret Naval experimental project involving a destroyer escort vessel and a process which he could identify only as "degaussing." On board the vessel, Silverman noted that there was "enough radar equipment on the ship to fill a battleship" including "an extra mast" which was "rigged out like a Christmas tree" with what appeared to be antenna-like structures.
>
> At one point during the preparation for the experiment, Silverman remembers seeing a civilian on board and said to a shipmate: "That guy could use a haircut." To his amazement he later discovered that the man had been Albert Einstein.
>
> Silverman was given the rating of Engineer, First Class, and, according to his account, was one of three seamen who knew where the switches were that started the operation. He also related that a special series of electrical cables had been laid from a nearby power house to the ship. When the order was given and the switches thrown, "the resulting whine was almost unbearable."[7]

One can only wonder if these experiments continued throughout the 1940s, 50s, 60s and even up to now. One might think that this is the case.

A Time Travel Reunion

As a final note on the history of the Philadelphia Experiment, it was reported in the *Bucks Courier Times* on Wednesday, March 24, 1999 that a time travel reunion of survivors of the Philadelphia Experiment would be held sometime that week.

Said the headline: "'Invisible' Ship Crew to Hold Reunion: About 15 members of a ship made famous by a supposed experiment in invisibility at the Philadelphia Navy Yard during World War II are gathering for the first time in 53 years."

No follow-up article has been reported, so we don't know whether the legend of the Philadelphia Experiment has been put to rest or not. Probably not. Myths and legends die hard, especially ones that span different periods of "time" as the Philadelphia Experiment and the Montauk Project have been claimed to do.

Said one researcher into the whole mess, "It will all be exposed and uncovered—in due time."

DEPARTMENT OF THE NAVY
OFFICE OF INFORMATION
WASHINGTON, D.C. 20350

IN REPLY REFER TO
OI-2252A/JVV/dh

2 3 JUL 1976

Over the years we have received innumerable queries about the so-called "Philadelphia Experiment" or "Project" and the alleged role of the Office of Naval Research (ONR) in it. The frequency of these queries predictably intensifies each time the experiment is mentioned by the popular press, often in a science fiction book.

The genesis of the Philadelphia Experiment myth dates back to 1955 with the publication of The Case for UFO's by the late Dr. Morris K. Jessup, a scientist with a Ph.D. in astrophysics and a varied career background.

Some time after the publication of the book, Dr. Jessup received a letter by a Carlos Miquel Allende, who gave his address as R.D. #1, Box 223, New Kensington, PA. In the letter, Allende commented on Dr. Jessup's book and gave details of an alleged secret naval experiment in Philadelphia in 1943. During the experiment, according to Allende, a ship was rendered invisible and tele-ported to and from Norfolk in a few minutes, with some terrible after-effects for the crew members. Supposedly, this incredible feat was accomplished by applying Einstein's never-completed "unified field" theory. Allende claimed that he had witnessed the experiment from another ship and that the incident was reported in a Philadelphia newspaper. Neither the identify of Allende, nor that of the newspaper has ever been established.

In 1956 a copy of Jessup's book was mailed anonymously to Admiral Furth, the Chief of Naval Research. The pages of the book were interspersed with hand-written annotations and marginalia apparently made by three different persons as they passed the book back and forth among them. The notations implied a knowledge of UFO's, their means of motion and generally, the culture and ethos of the beings occupying these UFO's.

The book came to the attention of two officers then assigned to ONR who happened to have a personal interest in the subject. It was they who contacted Dr. Jessup and asked him to take a look at his book. By the wording and style of one of the writers of the notations, Dr. Jessup concluded that the writer was the

The form letter sent to by the Department of the Navy to persons requesting information about the Philadelphia Experiment. This one from 1974, page 1.

same person who had written him about the Philadelphia Experiment. It was also these two officers who personally had the book retyped and who arranged for the publication, in typewritten form, of 25 copies. The officers and their personal belongings have left ONR many years ago, and we do not have even a file copy of the annotated book.

The Office of Naval Research never conducted an official study of the manuscript. As for the Philadelphia Experiment itself, ONR has never conducted any investigations on invisibility, either in 1943 or at any other time. (ONR was established in 1946.) In view of present scientific knowledge, our scientists do not believe that such an experiment could be possible except in the realm of science fiction. A scientific discovery of such import, if it had in fact occurred, could hardly remain secret for such a long time.

I hope this provides a satisfactory answer to your inquiry.

Sincerely,

Betty W. Shirley

BETTY W. SHIRLEY
Head, Research and Public
Inquiries Section

The form letter sent to by the Department of the Navy to persons requesting information about the Philadelphia Experiment. This one from 1974, page 2.

Chapter 6

Wernher von Braun and DISC

There's Nazis in the bathroom just below the stairs
Always something happening and nothing going on...
There's UFOs over New York and I ain't too surprised
—John Lennon, *Nobody Told Me*

In 1970 a photocopied manuscript began circulating among conspiracy researchers entitled "Nomenclature of an Assassination Cabal" by William Torbitt (a pseudonym). It became known as "The Torbitt Document" and was a damning expose of J. Edgar Hoover, Lyndon Johnson, John Connally and Wernher von Braun, among many others, for their role in the assassination of President John F. Kennedy. The manuscript was later published by Adventures Unlimited Press in 1996 as a paperback book entitled *NASA, Nazis and JFK*.[19] The late Kenn Thomsas of Steamshovel Press compiled the book and wrote an introduction.

This paperback edition of the Torbitt Document emphasizes what the manuscript says about the link between Operation Paperclip Nazi scientists working for NASA, an organization called the Defense Industrial Security Command (DISC), the assassination of JFK, and the secret Nevada air base known as Area 51. The Torbitt Document illuminates the darker side of NASA, the Military Industrial Complex, and the connections to Mercury, Nevada, and the Area 51 complex that headquarters the "secret space program."

When the paper was published, the author, Torbitt, claimed he was a lawyer working in the southwestern part of the United States.

The late Jim Marrs, when he was working for the *Fort Worth Star-Telegram,* in the early 1970s interviewed Torbitt and said that his real name was David Copeland, a lawyer from Waco, Texas.

Copeland allegedly told Marrs, "he got his information from two government agents—one with the Secret Service and the other with the FBI."

During the Second World War Copeland/Torbitt served in the United States Navy. After completing a law degree from the University of Texas he worked as a prosecuting attorney from 1949 to 1951. He admitted that his clients included people involved in committing political murder. He claimed he had also represented people involved in the "financial dealings of organized crime in Texas."

In "Nomenclature of an Assassination Cabal" Torbitt claims that John F. Kennedy was assassinated by a "fascist cabal... who planned to lay the blame on honest right-wing conservatives, if their first ploy, to lay the blame on Oswald and the Communists, was not bought."

Torbitt argues that a Swiss Corporation named Permindex engineered the assassination. Also involved was an organization called the Defense Industrial Security Command, organized by J. Edgar Hoover and William Sullivan of Division Five of the FBI. Torbitt claims that DISC agents included Clay Shaw, Guy Banister, David Ferrie, Lee Harvey Oswald, and Jack Ruby with Louis M. Bloomfield of Montreal, Canada in charge.

According to the author Permindex was comprised of:

(1) Solidarists, an Eastern European exile organization.

(2) American Council of Christian Churches led by Haroldson L. Hunt.

(3) Free Cuba Committee headed by Carlos Prio Socarras.

(4) The Syndicate headed by Clifford Jones, ex-lieutenant governor of Nevada. This group also included Bobby Baker, George Smathers, Roy Cohn, Fred Black and Lewis McWillie.

(5) Security Division of NASA headed by Wernher von Braun.

According to Torbitt, others involved in the assassination included Lyndon B. Johnson, John Connally, William Seymour, Robert McKeown, Sergio Arcacha Smith, Walter Jenkins, Fred Korth, Lee Harvey Oswald, Ruth Paine, Michael Paine, Gordon Novel, and Clint Murchison.

Torbitt adds that the "anti-Castro Cuban part of the plan was to tie the Castro regime into the murder of Kennedy and thus to

Clay Shaw.

have the US military give all service to the overthrow of Castro."

Let us take a quick look at two of these people: Louis Mortimer Bloomfield, said to be the main organizer, and William Sullivan of the FBI.

Louis Mortimer Bloomfield (born August 8, 1906) was a Canadian lawyer, businessman, and soldier. Bloomfield was recognized as a leader of the Canadian Jewish community for many decades.

Bloomfield earned a Bachelor of Arts from McGill University in 1927 and a Master of Laws from the University of Montreal

107

in 1930. He also received a Doctor of Laws from St. Francis Xavier University in 1964 and a Doctor of Civil Law from St. Thomas University in 1973.

Louis M. Bloomfield.

Bloomfield was admitted to the Bar of Quebec in 1930 and practiced international law with Phillips, Bloomfield, Vineberg, and Goodman from 1930 to 1970 in Montreal. Bloomfield assisted King Carol II of Romania in his attempt to gain entry into Canada after World War II. He was appointed a King's Counsel in 1948 and was elected to serve on the Mixed Court of Tangier (Tribunal Mixte Tangier) within the international zone of Tangier in 1949. At this point he became involved with French and Italian politics as well as Zionism.

In 1952, Bloomfield co-founded the Canadian branch of the International Law Association with Maxwell Cohen, Gerald F. FitzGerald, and Nicolas Mateesco-Matte. He served as that organization's president from 1964 to 1978, and was an honorary president from 1974 until his death in 1984. During this time he was a major figure in Canadian politics and law.

Proponents of some John F. Kennedy assassination conspiracy theories, including Torbitt/Copeland, have alleged that Bloomfield was linked to the Kennedy assassination through the Office of Strategic Services, the Central Intelligence Agency, and the Swiss registered company Permindex. Says Wikipedia about Bloomfield:

> On March 4, 1967, the Italian left-wing newspaper *Paese Sera* published a story alleging that Clay Shaw, who was arrested and charged with conspiring to assassinate President John F. Kennedy by New Orleans District Attorney Jim Garrison three days earlier, was linked to the CIA through his involvement in the Centro Mondiale Commerciale, a subsidiary of Permindex in which Shaw was a board member. According to *Paese Sera*, the CMC had been a front organization developed by the CIA for transferring funds to Italy for "illegal political-espionage activities" and had attempted to depose French President

108

Charles de Gaulle in the early 1960s. On March 6, the newspaper printed other allegations about individuals it said were connected to Permindex, including Bloomfield whom it described as "an American agent who now plays the role of a businessman from Canada [who] established secret ties in Rome with Deputies of the Christian Democrats and neo-Fascist parties."

So we see that there is some connection with Bloomfield and Permindex and Italian politics plus the attempted assassination of Charles de Gaulle, who pulled France out of NATO after the attempt.

William Sullivan and Division Five

William Cornelius Sullivan (May 12, 1912—November 9, 1977) was an assistant director of the Federal Bureau of Investigation who was in charge of the agency's domestic intelligence operations from 1961 to 1971. Sullivan became an FBI agent in 1941 and was transferred to El Paso, Texas in 1942.

Sullivan led the highly controversial COINTELPRO aimed at surveilling, infiltrating, discrediting, and disrupting domestic American political organizations, political opposition and civil rights movements. Many of these organizations and individuals were imprisoned, publicly humiliated, assassinated, or falsely charged with crimes.

According to Torbitt/Copeland Sullivan was the head of the secretive "Division Five" of the FBI that included dirty tricks and arranged assassinations among other things. Some of the Men in Black that have been talked about in UFO groups for decades may have come from Division Five, as well as from von Braun's DISC commanded from Huntsville.

Sullivan was forced out of the FBI at the end of September 1971 due to disagreements with FBI director J. Edgar Hoover. This is one year after the Torbitt Document began circulating. Sullivan was specifically mentioned in the document and one wonders if this had anything to do with his disagreements with Hoover.

The following year, in June 1972 Sullivan was appointed as the head of the Justice Department's new Office of National

William Sullivan, Deputy Director of the FBI.

Narcotics Intelligence, which he led until July 1973. Sullivan died in a suspicious hunting accident in 1977.

Sullivan claimed Hoover's concerns about the American Communist Party were overemphasized when compared to violations of federal civil rights laws in the segregated South. Many FBI insiders considered Sullivan the logical successor to Hoover. However, on October 1, 1971, Hoover abruptly had the locks changed on Sullivan's door and removed his nameplate. Sullivan was essentially forced to retire. Hoover would die less than a year later.

Sullivan then became even more vocal about Hoover's controversial domestic counterintelligence programs, collectively labeled COINTELPRO, including operations that he himself had conceived and administered. These were intended to spread confusion and dissension among political groups in the United States ranging from the Civil Rights Movement, the Communist Party (CPUSA), and the anti–Vietnam War movement on the left to the Ku Klux Klan on the far right.

Sullivan described the typical frame-up of Communist Party leader William Albertson in a June 30, 1964, internal document filed by him and fellow FBI agent Fred J. Baumgardner:

> My memorandum dated 6/12/64 was approved, authorizing a unique counterintelligence operation calculated to cast suspicion on Communist Party (CP) National Committee member and Executive Secretary of the New York District Organization, William Albertson. It was our intention to place Albertson in the unenviable position of being suspected as an FBI informant through the use of a planted bogus informant report prepared by the Laboratory in Albertson's handwriting on paper used by him with a ballpoint pen of the type he uses.

Sullivan was instrumental in arranging for the mailing of a tape recording in 1964 to Coretta Scott King which contained secretly taped recordings of her husband Martin Luther King Jr. having relations with other women. In a memorandum, Sullivan called King "a fraud, demagogue and scoundrel." He also gave orders to track down fugitive members of the Weather Underground in the early 1970s.

Sullivan personally handled the FBI's investigations into the assassinations of JFK, RFK and Martin Luther King. He found no wrongdoing on the part of any government agency and firmly announced that each assassination was the work of a lone nut assassin.

After Hoover's death in May 1972, U.S. Attorney General Richard Kleindienst appointed Sullivan director of the newly created Office of National Narcotics Intelligence under the Department of Justice in June 1972. Sullivan had hoped to replace Hoover as the director of the FBI, but was passed over by President Richard Nixon in favor of loyalist L. Patrick Gray.

Sullivan and Hoover.

In 1975 Sullivan was called before the Senate Intelligence Committee and was asked if he had ever seen anything in the FBI files to indicate a relationship between Oswald and the CIA. Sullivan answered cryptically:

> No... I think there may be something on that, but you asked me if I had seen anything. I don't recall having seen anything like that, but I think there is something on that point... it rings a bell in my head.

Sullivan also testified before the 1975 Senate Intelligence Committee: "Never once did I hear anybody, including myself raise the question, is this course of action which we have agreed upon lawful, is it legal, is it ethical or moral?" He seemed to be having moral problems with things he had done in the past which did not just include the defamation of people by false means but included assassinations.

Things were not going well for Sullivan as to the questions he was being asked by the Senate Intelligence Committee; and they wanted to ask him more questions in 1977, but before he could testify he was shot to death in a "hunting accident" on November 9, 1977 at the age of 65.

Recounted by the journalist Robert D. Novak, who says that Sullivan predicted his death:

> Sullivan came to our house in the Maryland suburbs in June 1972 for lunch and a long conversation about my plans for a biography of Hoover (a project I abandoned as just too ambitious an undertaking). Before he left, Bill told me someday I probably would read about his death in some kind of accident, but not to believe it. It would be murder.

On November 9, 1977, days before he was to testify to the House Select Committee on Assassinations, twenty minutes before sunrise, sixty-five-year-old Sullivan was walking through the woods near his retirement home in Sugar Hill, New Hampshire, on the way to meet hunting companions. Another hunter, Robert Daniels, Jr., a twenty-two-year-old son of a state policeman, using a telescopic sight on a .30 caliber rifle, said he mistook Sullivan for a deer, shot him in the neck, and killed him instantly.

The authorities called it an accident, fining Daniels five hundred dollars and taking away his hunting license for ten years. Sullivan's death did not prevent the 1979 publication of a memoir, telling all about the disgrace of J. Edgar Hoover and the FBI. After Watergate, with all the principals dead or out of office, the book received little attention.

His memoir of his thirty-year career in the FBI, written with journalist Bill Brown, was published posthumously by commercial publisher W. W. Norton & Company in 1979. It seems that William Sullivan just knew too much to keep talking to Senate committees. Once again Torbitt/Copeland seems to have something there when he names these people. Ultimately there are just too many people to examine who are named in the document.

In his book *Who Shot JFK?*[54] British JFK expert Robin Ramsay argues that "Nomenclature of an Assassination" Cabal was an attempt by the Central Intelligence Agency to link the Federal Bureau of Investigation to the assassination of John F. Kennedy. He says:

> Torbitt took Garrison's inquiry into the CIA's links to the assassination and converted them into a story about the FBI's responsibility for the assassination. (This, in my view, tells us that the author/s of Torbitt were working for the CIA, trying to diminish the "Garrison effect.")

This "Garrison effect" was that Garrison was seemingly finding a CIA connection with Oswald, Clay Shaw and others. Was the CIA involved? If Oswald was a CIA agent then he was also a

113

patsy, as he said himself shortly before being shot at the Dallas police station by Jack Ruby. Maybe the CIA wasn't so involved in the actual assassination, except for the Cuban operatives, but was busy with the cover-up. If LBJ was involved in the assassination as Torbitt/Copeland allege then a massive cover-up would have to occur—a cover-up that would include murdering dozens of people, starting with apparent CIA operative Lee Harvey Oswald.

As for Sullivan, on the day of the JFK assassination he was in Washington DC. By 6 pm that day Sullivan was in charge of investigating the chief suspect in the assassination: Lee Harvey Oswald. Sullivan was possibly just as surprised as everyone that Kennedy had been shot, but he was now part of the massive cover-up that was about to occur. Did he know that Wernher von Braun was somehow involved? Indeed, while the CIA and the FBI began pointing fingers at each other perhaps there was a third power involved: that of Wernher von Braun, Paperclip Nazis and DISC.

DISC and the Defense Counterintelligence and Security Agency

If one does an Internet search for DISC or the Defense Industrial Security Command the only thing that comes up is references to DISC in the Torbitt Document. DISC does not seem to exist otherwise. It is only found in the Torbitt Document which says that Hoover, Sullivan and Wernher von Braun were part of DISC. No mention of DISC can be found in any of their biographies.

However, what does come up in an Internet search is the Defense Counterintelligence and Security Agency or DCSA. The US government website for the DCSA says that the organization was founded in 1971, one year after the Torbitt Document was first released. According to its website the mission of the DCSA is:

> Through vetting, industry engagement, education, and counterintelligence and insider threat support, secure the trustworthiness of the United States Government's workforce, the integrity of its cleared contractor support, and the uncompromised nature of its technologies, services, and supply chains.
> Our Vision

114

DCSA is America's Gatekeeper: Safeguarding the nation as the premier provider of integrated security services—national security is our mission, people are our greatest asset.

Under a page titled "History" the website says:

The history of DCSA traces its origins to two key missions—personnel security and industrial security. Over the past 50 years, the agency has been defined and redefined by these linked core mission sets.

Our Past Directors:
Brigadier General Joseph Cappucci, USAF, 1971-1976
Bernard J. O'Donnell, 1976 - 1981
Thomas J. O'Brien, 1981 - 1988
John F. Donnelly, 1988 - 1996
Margaret R. Munson, 1996 - 1998
Steven T. Schanzer, 1998 - 1999
Charles J. Cunningham Jr., 1999 - 2002
William Curtis, 2002 - 2004
Heather Anderson, 2004 - 2005
Janice Haith, 2005 - 2006
Kathy Watson, 2006 - 2010
Stanley L. Sims, 2010 - 2016
Daniel E. Payne, 2016 - 2019
Charles Phalen 2019 - 2020
William K. Lietzau 2020 - 2023
Daniel J. Lecce 2023 - 2024
David M. Cattler 2024 - present

Mission History
Personnel security and industrial security are the core missions of our agency. Over the past 50 years, the DCSA has been defined and redefined by these missions: Personnel Vetting, Industrial Security, Education and Training
On December 29, 1971, the Secretary of Defense established the Defense Investigative Service (DIS).

This began the Department's unified handling of its personnel security, effective January 1, 1972. DIS's tasks, responsibilities, and authority were published in DoD Directive 5105.42 and designated DIS as a separate operating agency under the direction of the Secretary.

Realignment to OPM

In 1999, DIS underwent a reorganization to become the Defense Security Service (DSS). DSS retained the personnel security investigation (PSI) mission until February 20, 2005, when the function was transferred to the Office of Personnel Management (OPM). This transfer included PSIs for industry personnel under the National Industrial Security Program. It also included the transfer of roughly 1,850 personnel, which was stipulated in the National Defense Authorization Act for Fiscal Year 2004.

OPM, as the successor to the Civil Service Commission, has been conducting PSIs for most non-DoD programs since 1953. That is when EO 10450 established the requirement for a government-wide PSI program and granted authority for the program to the Civil Service Commission. OPM became the government's largest ISP in 2005 after assuming the DSS program and personnel.

Realignment to DCSA

In October 2016, the semi-autonomous National Background Investigations Bureau (NBIB) was established under OPM. NBIB was the primary investigative service provider (ISP) for the Federal Government. It conducted ~95% of all federal background investigations. It began the process of transforming the PSI mission. NBIB kept that mission until the program and its personnel were moved to DCSA.

On April 24, 2019, Executive Order (EO) 13869 directed the transfer of NBIB from OPM to the Department of Defense (DOD), effective October 1, 2019.

So, the DCSA was founded by the DoD in 1971 and its first director was Brigadier General Joseph Cappucci, USAF, 1971-1976. Cappucci was a career military officer who first joined the

US Army Air Corps. He was a high-ranking and well-traveled Air Force officer. Wikipedia has something interesting to say about Cappucci and the Kennedy assassination:

> Joseph J. Cappucci (1 January 1913—10 June 1992) was a U.S. Air Force brigadier general (Special Agent) who served as the first director of the Defense Investigative Service and the 6th Commander of the Air Force Office of Special Investigations (AFOSI or OSI). As the Defense

Brigadier General Joseph J. Cappucci.

Investigative Service director, he oversaw its tasks of facilitating personnel security investigations, supervising industrial security, and performing security education and awareness training.

...In the late 1960s, Cappucci was commander of the 1005th Special Investigations Group, which gave him access to many individuals in the intelligence community including a close and long-term relationship with former FBI director J. Edgar Hoover. During a dinner at a hotel in Rome, Italy, with Lieutenant Colonel William H. Amos and his wife, Cappuci talked at length about inside knowledge of Mary Jo Kopechne's death inside of former Senator Ted Kennedy's car at Chappaquiddick Island, MA. It became clear to Mrs. Amos, Cappucci obtained inside information about the investigation of the accident from Hoover and how Ms. Kopechne's parents agreed to a settlement with Senator Kennedy to remain silent. Afterwards, the conversation turned to former President John F. Kennedy's assassination wherein Cappucci told the couple, "it was no wonder LBJ had JFK killed." After the dinner, Lt Col Amos told his wife to never repeat what Cappucci said, which she did for nearly 50 years.

This claim that LBJ was part of the conspiracy to kill Kennedy is part of the Torbitt Document and Cappucci seems to be giving the document some credibility. I maintain that more than anything it is that LBJ was part of the conspiracy that the CIA and the FBI had to hide that fact no matter what. That is why all the JFK files have not been released—because they talk about LBJ and his cohorts.

The KGB concluded that Johnson was part of the assassination plot as did apparently British Intelligence. While British Intelligence has never publicly weighed in on the Kennedy assassination some of the top JFK assassination authors are British and the UK has published numerous books on the subject. It is assumed that much of the excellent information in these many books comes from contacts in British Intelligence. It might be noted here that Queen Elizabeth did not meet with LBJ, the only US president that she passed over.

But what really concerns us here is what kind of organization existed before the creation of the DCSA. Certainly there must have been some organization that essentially vetted the Defense Industrial Complex that President Eisenhower famously warned us about in 1959. Was it DISC, being run by Hoover, Sullivan and Wernher von Braun from Huntsville, Alabama? The government website for the DCSA gives Huntsville as one of the Field Offices. The headquarters of the DCSA is in Quantico, Virginia, the same city that houses the headquarters of the FBI.

It would seem that the DCSA is the final incarnation of Torbitt's DISC. Why should we think that Torbitt completely fabricated the organization called DISC? Some such organization must have existed within the military and FBI, and if it was not called DISC then it was probably called something similar.

Torbitt specifically says that Wernher von Braun commanded DISC from his office in Huntsville, Alabama. Why would Torbitt fabricate both DISC and von Braun's official position within the organization? Indeed, there seems to be something to von Braun's involvement with DISC as well as in the alleged murder of Morris K. Jessup.

Who was Wernher von Braun?

Wernher Magnus Maximilian Freiherr von Braun was born on March 23. 1912. He died on June 16, 1977. He was a German aerospace engineer and space architect. He was a member of the Nazi Party and Allgemeine SS, the leading figure in the development of rocket technology in Nazi Germany, and later a pioneer of rocket and space technology in the United States. He was essentially the head of NASA starting in 1960.

As a young man, von Braun worked in Nazi Germany's rocket development program. He helped design and co-developed the V-2 rocket at Peenemünde during World War II. The V-2 became the first artificial object to travel into space on June 20, 1944. Following the war, he was secretly moved to the United States, along with about 1,600 other German scientists, engineers, and technicians, as part of Operation Paperclip. He worked for the United States Army on an intermediate-range ballistic missile

Wernher von Braun and Rudolf Nebel with early rockets, circa 1932.

program, and he developed the rockets that launched the United States' first space satellite Explorer 1 in 1958.

Von Braun is a highly controversial figure widely seen as escaping justice for his Nazi war crimes due to the Americans' desire to beat the Soviets in the Cold War. He is also sometimes described by others as the "father of space travel," the "father of rocket science," or the "father of the American lunar program." He advocated a human mission to Mars.

Von Braun joined the SS horseback riding school in 1933 as an SS-Anwärter. He left the following year. In 1940, von Braun joined the SS and was given the rank of Untersturmführer in the Allgemeine-SS and issued a membership number. In 1947, he gave the US War Department this explanation:

> In spring 1940, one SS-Standartenführer (SS-Colonel) Müller from Greifswald, a bigger town in the vicinity of Peenemünde, looked me up in my office... and told me that Reichsführer-SS Himmler had sent him with the order to urge me to join the SS. I told him I was so busy with my rocket work that I had no time to spare for any political activity. He then told me, that... the SS would cost me no time at all. I would be awarded the rank of a[n] "Untersturmfuehrer" (lieutenant) and it were [sic] a very definite desire of Himmler that I attend his invitation to join.
>
> I asked Müller to give me some time for reflection. He

120

agreed.

Realizing that the matter was of highly political significance for the relation between the SS and the Army, I called immediately on my military superior, Dr. Dornberger. He informed me that the SS had for a long time been trying to get their "finger in the pie" of the rocket work. I asked him what to do. He replied on the spot that if I wanted to continue our mutual work, I had no alternative but to join.

When shown a picture of himself standing behind Himmler, von Braun said that he had only worn the SS uniform that one time, but Wikipedia says that in 2002 a former SS officer at Peenemünde told the BBC that von Braun had regularly worn the SS uniform to official meetings. He began as an Untersturmführer (Second lieutenant) and was promoted three times by Himmler, the last time in June 1943 to SS-Sturmbannführer (Major). Von Braun later stated that these were simply technical promotions received each year regularly by mail.

In 1933, von Braun was working on his creative doctorate when the Nazi Party came to power in a coalition government in Germany; rocketry was almost immediately moved onto the national agenda. An artillery captain, Walter Dornberger, arranged an Ordnance Department research grant for von Braun, who then worked next to Dornberger at the solid-fuel rocket test site at

Wernher von Braun at Peenemünde with Emil Leeb and Fritz Todt.

Kummersdorf, Germany.

Von Braun received his doctorate in physics (aerospace engineering) on July 27, 1934 from the University of Berlin with a thesis titled "Construction, Theoretical, and Experimental Solution to the Problem of the Liquid Propellant Rocket" that detailed the construction and design of the A2 rocket. It remained classified by the German army until its publication in 1960. By the end of 1934, his group had successfully launched two liquid fuel A2 rockets. Von Braun continued his guided missile work throughout World War II, and met with Adolf Hitler on several occasions, being formally decorated by Hitler twice, including being awarded the Iron Cross.

On December 22, 1942, Hitler ordered the production of the A-4 as a "vengeance weapon," and the Peenemünde group developed it to target London. Following von Braun's July 7, 1943 presentation of a color movie showing an A-4 taking off, Hitler was so enthusiastic that he personally made von Braun a professor shortly thereafter.

By that time, the British and Soviet intelligence agencies were aware of the rocket program and von Braun's team at Peenemünde, based on the intelligence provided by the Polish underground Home Army. Over the nights of 17–18 August 1943, RAF Bomber Command's Operation Hydra dispatched raids on the Peenemünde camp consisting of 596 aircraft, and dropped 1,800 tons of explosives.

The facility was rebuilt and most of the engineering team remained unharmed; however, the raids killed von Braun's engine designer Walter Thiel and Chief Engineer Walther, and the rocket program was delayed. On June 20, 1944 the V-2 became the first artificial object to travel into space by crossing the Kármán line with the vertical launch of rocket MW 18014.

The first combat A-4 was

A V-2 launch from Peenemünde.

renamed the V-2 again which was thought advantageous for propaganda purposes because V-2 stood for "Vergeltungswaffe 2" or "Retaliation/Vengeance Weapon 2." The first of these A-4/V-2 rockets was launched toward England on September 8, 1944. It was only 21 months after the project was officially commissioned that this first attack took place.

Wikipedia says that Doug Millard of the Science Museum, London states:

> The V-2 was a quantum leap of technological change. We got to the Moon using V-2 technology but this was technology that was developed with massive resources, including some particularly grim ones. The V-2 programme was hugely expensive in terms of lives, with the Nazis using slave labor to manufacture these rockets.

Enter SS General Hans Kammler, who as an engineer had constructed several concentration camps, including Auschwitz, and had a reputation for brutality. Kammler had conceived the idea of using concentration camp prisoners as slave laborers in the rocket program. Arthur Rudolph, chief engineer of the V-2 rocket factory at Peenemünde, endorsed this idea in April 1943 when a labor shortage developed. More people died building the V-2 rockets than were killed by it as a weapon. Von Braun admitted visiting the plant at Mittelwerk on many occasions, and called conditions at the plant "repulsive," but stated that he had never personally witnessed any deaths or beatings, although it had become clear to him by 1944 that deaths had occurred. He denied ever having visited the Mittelbau-Dora concentration camp, where 20,000 died from illness, beatings, hangings, and intolerable working conditions.

The Mysterious Hans Kammler

Hans Kammler (26 August 1901 – 1945 [assumed]) was an SS-Obergruppenführer responsible for Nazi civil engineering projects and its top secret V-weapons program. He oversaw the construction of various Nazi concentration camps before being put in charge of the V-2 rocket and Emergency Fighter programs towards the end of World War II. Kammler disappeared in May 1945 during the final days of the war.

Kammler was also involved in an attempt to finish the Blockhaus d'Éperlecques near Saint-Omer, Pas-de-Calais in Northern France. The fortified bunker was to be used as a V-2 launch base but it was abandoned in September 1944 before it was finished.

SS General Hans Kammler.

In March 1944, Himmler convinced Adolf Hitler to put the V-2 project directly under SS control. On 8 August Kammler replaced Walter Dornberger as its director taking command and control of V-2 operations. The first rockets were launched from a site near The Hague against London on September 8, 1944. On January 31, 1945, Hitler named Kammler head of all missile projects, but by this time the lack of explosives was critical and the program was winding down. Germany was losing the war and had a severe lack of resources.

In March 1945, Hitler stripped Göring of his powers over aircraft support, maintenance and supply, and transferred all of these duties to Kammler, who was an SS officer of the highest rank. This meant that in the beginning of April 1945, just weeks before the official surrender, Kammler was raised to "Fuehrer's general plenipotentiary for jet aircraft." He was the commander of all aircraft in the Reich which was a combination of SS aircraft and the Wehrmacht aircraft of the conventional German Army.

Kammler now had numerous aircraft at his disposal—including captured British and American aircraft. He also had control of all the resources of the secret extra-territorial bases that the Germans had and were now controlled by the SS. This included submarines and the secret submarine bases in Antarctica, Tierra del Fuego, Greenland and the Canary Islands and more. Kammler was essentially in control of the postwar Reich which would largely be centered in South America and briefly Antarctica.

In the final weeks of the war in Europe, Kammler's movements became sketchy and contradictory. Wernher von Braun said

124

Kammler was in Oberammergau, Bavaria, in April 1945. The Nazi rocket scientist later reported having overheard a discussion between Kammler and his aide-de-camp in which Kammler said he planned to hide in nearby Ettal Abbey. Kammler and his followers then left town, according to von Braun.

Further evidence of Kammler's activities consists of a telegraph sent from Kammler to Speer, Himmler, and Göring on April 16, informing them of the creation of a "message center" at Munich and the appointment of an operations chief for the construction of the Messerschmitt Me 262. On April 20, he reportedly arrived with a group of technicians at Himmler's Kommandostelle near Salzburg. On April 23, Kammler sent a radio message to his office manager at Berlin, ordering him to organize the immediate destruction of the "V-1 equipment near Berlin" and then to go to Munich.

Wikipedia says that the Combined Intelligence Objectives Sub-Committee (CIOS) in London ordered a search for Kammler in early July 1945. Members of the 12th Army replied that he was last seen on 8 or 9 April in the Harz region. In August, Kammler's name made "List 13" of the UN for Nazi war criminals. Only in 1948 did the CIOS receive the information that Kammler reportedly fled to Prague and had committed suicide.

Wikipedia says an Army Counter Intelligence Corps (CIC) report from April 1946 listed Kammler among SS officers known to be outside Germany and considered to be of special interest to the CIC. In mid-July 1945, the head of the Gmunden CIC office, Major Morrisson, interviewed an unnamed German on the issue of a numbered account associated with construction sites for plane

SS General Hans Kammler.

and missile production formerly run by the SS. A report published years later, in late 1947 or early 1948, stated that only Kammler and two other persons had access to the account.

Kammler is thought to have disappeared in Prague in May of 1945. Claims were made of a suicide by cynanide, but no body was ever found nor any reports of disposing of the body. Some claimed to have seen him after the war. It has been speculated that he used the "Ratline" to escape to South America via Italy and Spain. It has been said that the Americans captured Kammler and also that he escaped from them. The US military says they have no information on Kammler.

Let us not forget that Kammler was the commander of the postwar aircraft that had not surrendered. He had numerous aircraft to fly around in, including long-range experimental aircraft and even presumably the Vril and Haunebu flying saucer aircraft being developed and test flown.

Kammler simply vanished at the end of the war. He likely travelled to nearly all of the secret bases that the Nazis and SS had created before and during the war. He was a very important engineer and SS operative who knew Wernher von Braun and Walter Dornberger, both of whom went on to work for the US Army and NASA. Is it possible that Hans Kammler was somehow part of Operation Paperclip that brought Nazi scientists to the United States? Kammler remains the greatest mystery in regards to Operation Paperclip.

Wernher von Braun Escapes the Third Reich

Wikipedia says that according to André Sellier, a French historian and survivor of the Mittelbau-Dora concentration camp, Heinrich Himmler had von Braun come to his Feldkommandostelle Hochwald HQ in East Prussia in February 1944. To increase his power base within the Nazi regime, Himmler was conspiring to use Hans Kammler to gain control of all German armament programs, including the V-2 rocket program at Peenemünde. Himmler therefore recommended that von Braun work more closely with Kammler to solve the problems of the V-2. Von Braun stated that he replied that the problems were merely technical and he was confident that they would be solved with Dornberger's assistance.

Von Braun had been under SD (a branch of the SS) surveillance since October 1943. A secret report stated that he and his colleagues Klaus Riedel and Helmut Gröttrup were said to have expressed regret at an engineer's house one evening in early March 1944 that they were not working on a spaceship and that they felt the war was not going well; this was considered a "defeatist" attitude. A young female dentist who was an SS spy reported their comments.

Himmler's unfounded allegations branded von Braun and his colleagues as communist sympathizers and accused them of sabotaging the V-2 program. This was considered a threat, coupled with von Braun's regular piloting of a government-provided airplane that could facilitate an escape to Britain.

The unsuspecting von Braun was detained on March 14, 1944, and was taken to a Gestapo cell in Szczecin, Poland. Here he was held for two weeks without knowing the charges against him.

Wikipedia tells us that through Major Hans Georg Klamroth, in charge of the Abwehr for Peenemünde, Dornberger obtained von Braun's conditional release. Albert Speer, Reichsminister for Munitions and War Production, persuaded Hitler to reinstate von Braun so that the V-2 program could continue and ultimately be called the "V-4 program" which in their view would be impossible without von Braun's leadership. In his memoirs, Speer states Hitler had finally conceded that von Braun was to be "protected from all prosecution as long as he is indispensable, difficult though the general consequences arising from the situation."

In early 1945 the Soviet Army was about 160 km (100 miles) from Peenemünde when von Braun assembled his planning staff and asked them to decide how and to whom they should surrender. Unwilling to go to the Soviets, von Braun and his staff decided to try to surrender to the Americans. Kammler had ordered the relocation of his team to central Germany; however, a conflicting order from an army chief commanded them to join the army and fight.

Deciding that Kammler's order was their best bet to defect to the Americans, von Braun fabricated documents and transported 500 of his affiliates to the area around Mittelwerk, where they resumed their work in Bleicherode and surrounding towns after the middle of February 1945. For fear of their documents being

127

destroyed by the SS, von Braun ordered the blueprints to be hidden in an abandoned iron mine in the Harz mountain range near Goslar. The Third Reich was essentially in chaos by this point.

While on an official trip in March, von Braun suffered a complicated fracture of his left arm and shoulder in a car accident after his driver fell asleep at the wheel. His injuries were serious, but he insisted that his arm be set in a cast so that he could leave the hospital. Due to this neglect of the injury, he had to be

Major General Walther Dornberger, Commander of the V-2 laboratory at Peenemunde with Lieutenant Colonel Herbert Axter and von Braun with the arm cast.

hospitalized again a month later when his bones had to be rebroken and realigned.

In early April, as the Allied forces advanced deeper into Germany, Kammler ordered the engineering team, around 450 specialists, to be moved by train into the town of Oberammergau in the Bavarian Alps, where they were closely guarded by the SS with orders to execute the team if they were about to fall into enemy hands. However, von Braun managed to convince SS Major Kummer to order the dispersal of the group into nearby villages so that they would not be an easy target for US bombers. On April 29, 1945, Oberammergau was captured by the Allied forces who seized the majority of the engineering team.

Nearing the end of the war, Hitler instructed SS troops to gas all technical men concerned with rocket development. Upon hearing this, von Braun commandeered a train and fled with other "technical men" to a location in the mountains of South Germany. After some time, von Braun and many of the others who made it to the mountains left their location to flee to advancing American lines in Austria.

Von Braun and several members of the engineering team, including Dornberger, made it to Austria. On May 2, 1945, upon finding an American private from the US 44th Infantry Division, von Braun's brother and fellow rocket engineer, Magnus, approached the soldier on a bicycle, calling out in broken English: "My name is Magnus von Braun. My brother invented the V-2. We want to surrender."

In September 1945, von Braun and other members of the Peenemünde team signed a work contract with the United States Army Ordnance Corps. On September 20, 1945, the first seven technicians arrived in the United States at New Castle Army Air Field, just south of Wilmington, Delaware. They were then flown to Boston, Massachusetts, and taken by boat to the Army Intelligence Service post at Fort Strong in Boston Harbor. Later, with the exception of von Braun, the men were transferred to Aberdeen Proving Ground in Maryland to sort out the Peenemünde documents, enabling the scientists to continue their rocketry experiments.

Finally, von Braun and his remaining Peenemünde staff

Wernher von Braun's identification card at Fort Bliss.

were transferred to their new home at Fort Bliss, a large Army installation just north of El Paso, Texas. Von Braun later wrote that he found it hard to develop a "genuine emotional attachment" to his new surroundings.

Wikipedia says that his lab was never able to get sufficient funds to go on with their programs. Von Braun remarked "at Peenemünde we had been coddled, here you were counting pennies." Whereas von Braun had thousands of engineers who answered to him at Peenemünde, he was now subordinate to "pimply" 26-year-old Jim Hamill, an Army major who possessed only an undergraduate degree in engineering. His loyal Germans still addressed him as "Herr Professor," but Hamill addressed him as "Wernher" and never responded to von Braun's request for more materials. Every proposal for new rocket ideas was dismissed.

While at Fort Bliss, von Braun and Dornberger trained military, industrial, and university personnel in the intricacies of rockets and guided missiles. As part of the Hermes project, they helped refurbish, assemble, and launch a number of V-2s that had been shipped from Allied-occupied Germany to the White Sands Proving Ground in New Mexico. They also continued to study the future potential of rockets for military and research applications.

In 1950, at the start of the Korean War, von Braun and his team were transferred to Huntsville, Alabama, his home for the next

20 years. From 1952 to 1956, von Braun led the Army's rocket development team at Redstone Arsenal, resulting in the Redstone rocket, which was used for the first live nuclear ballistic missile tests conducted by the United States. He personally witnessed this historic launch and detonation. Work on the Redstone led to the development of the first high-precision inertial guidance system on the Redstone rocket. By 1953 von Braun's title was "Chief, Guided Missiles Development Division, Redstone Arsenal."

As director of the Development Operations Division of the Army Ballistic Missile Agency, von Braun, with his team, then developed the Jupiter-C, a modified Redstone rocket. The Jupiter-C was the basis for the Juno I rocket that successfully launched the West's first satellite, Explorer 1, on January 31, 1958. This event was the birth of America's space program.

NASA was established on July 29, 1958. One day later, the 50th Redstone rocket was successfully launched from Johnston Atoll in the south Pacific as part of Operation Hardtack I. Two years later, NASA opened the Marshall Space Flight Center at Redstone Arsenal in Huntsville, and the Army Ballistic Missile Agency (ABMA) development team led by von Braun was transferred to NASA.

Wernher von Braun in his office at Redstone Arsenal in Huntsville, Alabama.

131

In a face-to-face meeting with Herb York at the Pentagon, von Braun made it clear he would go to NASA only if development of the Saturn were allowed to continue. Von Braun became NASA's first director on July 1, 1960 and held the position until January 27, 1970.

This is the period, 1958 to 1970, that von Braun apparently commanded DISC. He did it from the Redstone Arsenal in Huntsville and was in touch with the FBI on an ongoing basis. This is basically what the Torbitt Document says. It was from Redstone Arsenal that the "Men in Black" originated when they began a surveillance of Morris Jessup early on in the Varo Edition affair.

It has been surmised by this author and others that Jessup was killed by "Men in Black" from Huntsville. They would have probably killed Carl Allen as well, but they could not find him as he was an elusive "gypsy."

Von Braun was diagnosed in 1973 with kidney cancer during a routine medical examination. However, he continued to work unrestrained for a number of years. In 1977, President Gerald R. Ford awarded him the country's highest science honor, the National Medal of Science in Engineering but he was too ill to attend the White House ceremony. Von Braun died on June 16, 1977 of pancreatic cancer in Alexandria, Virginia at age 65.

And so we get to the end of a very interesting life. One that includes early rocket development, the dangers and adventures of WWII, and the American quest into space. Von Braun was a leading proponent of journeys to the Moon and Mars.

But what of DISC and von Braun? Did Torbitt just make this up? It seems that there is something to the story of DISC, Huntsville and von Braun. Were they involved in the murder of Morris K. Jessup? That is part of the mystery of Project Rainbow and the Philadelphia Experiment.

Chapter 7

The Einstein-Tesla Connection

I'm a thousand miles from nowhere
Time don't matter to me
'Cause I'm a thousand miles from nowhere
And there's no place I want to be
—*A Thousand Miles from Nowhere*, Dwight Yoakam

In the last chapter we looked into Wernher von Braun, Hans Kammler, Walter Dornberger and other Project Paperclip Nazis who either came to the US to work in the military rocket programs or escaped to South America and elsewhere. In my recent books on Nazi survival and secret technology such as *Antarctica and the Secret Space Program*[25] or *Vril: The Secret of the Black Sun*,[28] I have elucidated on published plans and photos of the Vril and Haunebu flying disks and the large Andromeda craft that was long and cylindrical. These craft were (are) electric and are powered by some form of Tesla technology.

The plans were first sent to a German who was living in London in 1989 and then published in the German language in Austria where the books were best sellers and were seen by a wide audience. I discuss all of this in detail in *Haunebu: The Secret Files*[26] and *Andromeda: The Secret Files*.[27] I include many photos, some in color, of these craft. This secret technology was developed by the SS during the war and development continued after the war at secret laboratory-bases in Antarctica, South America and Greenland.

Apparently a large number of black U-boats were still in service in the years immediately after the war. These "missing" submarines used bases in Greenland, Antarctica, South America (Tierra del Fuego), the Canary Islands, and probably the Spanish

133

territory of Western Sahara. It has also been stated that some of these black U-boats were used to bring Project Paperclip Nazis to Florida where they were either given new identities or were absorbed into the military industrial complex. Wernher von Braun is said to have had a hand in these former SS officers and scientists coming to the US.

Nikola Tesla was born in Serbia in 1865 and died during WWII on January 7, 1943. The Philadelphia Experiment supposedly happened in October later that year, specifically October 28, 1943. What was Tesla's contribution to Project Rainbow and the Philadelphia Experiment? We know that his inventions contributed to the Nazi flying saucers being built by the SS.

Similarly, what was Albert Einstein's contribution to the Philadelphia Experiment? Einstein was known to have had meetings with top officers of the US Navy. Carl Allen repeatedly spoke of Einstein and his Unified Field Theory as an important factor in the experiments that the Navy was conducting in 1943—whatever they were.

WWII was a time of stressful races to make better weapons and anti-weapons and faster aircraft and better submarines. The need for secrecy and speed was very important and large amounts of money were used for major experiments and projects. Project Rainbow and the Philadelphia Experiment was such an experiment. But what was it? Was it just a degaussing experiment or did it involve invisibility and teleportation? Was the US Navy actually helping William Moore and Charles Berlitz write their 1979 book?

Their next book was *The Roswell Incident*[29] published in 1980. It seems that by this time both authors were well connected with Navy Intelligence and CIA. But before their book *The Philadelphia Experiment* Charles Berlitz had written a book in 1977 called *Without a Trace*.[8] This was the first book published by a major publisher that discussed the Philadelphia Experiment. In that book Berlitz's friend and former intelligence operative J. Manson Valentine told Berlitz he had spoken to Jessup before his death:

> According to Jessup the purpose [of the Philadelphia Experiment] was to test out the effect of a strong magnetic

field on a manned surface craft. This was to be accomplished by means of magnetic generators (degaussers). Both pulsating and non-pulsating generators were operated...

Jessup was worried about the experiments and told Valentine that the Navy had requested him to be a consultant on yet another experiment but that he had refused. He was convinced that the Navy, in seeking to create a magnetic cloud for camouflage purposes in October 1943, had uncovered a potential that could temporarily, and if strong enough perhaps permanently, rearrange the molecular structure of people and materials so that they would pass into another dimension with further implications of predictable and as yet uncontrolled teleportation.

I do not believe Dr. Jessup considered this an 'inadvertent' discovery. For many years, so I have been told, experiments involving high intensity magnetism have been officially discouraged, just as the ion motors, known for as far back at least as 1918, have been denied public disclosure and their inventors somehow silenced. I am therefore convinced that top-ranking physicists must have some knowledge—and understandable dread—of phenomena that might be expected to emerge from the generation of a high intensity magnetic field, especially a pulsating or vortextual one.

QUESTION: In the case of the alleged Philadelphia Experiment, is there a fairly simple scientific explanation as to what took place?

ANSWER: To my knowledge there is no explanation in terms of the familiar or orthodox. Many scientists now share the opinion that basic atomic structure is essentially electric in nature rather than materially particulate. A vastly complicated interplay of energies is involved. Such a broad concept lends great flexibility to the universe. If multiple phases of matter within such a cosmos did NOT exist, it would be most surprising.

The transition from one phase to another would be equivalent to the passage from one plane of existence to another—a sort of interdimensional metamorphosis.

135

In other words, there could be 'worlds within worlds.' Magnetism has long been suspect as an involvement agent in such potentially drastic changes. To begin with, it happens to be the only inanimate phenomenon for which we have been unable to conceive a mechanistic analogue. We can visualize electrons traveling along a conductor and thus 'explain' electric current, or we can envisage energy waves of different frequencies in the ether and thus 'explain' the heat-light-radio spectrum. But a magnetic field defies a mechanical interpretation. There is something almost mystical about it. Furthermore, whenever we encounter incredible (to us) materialization and dematerialization, as in UFO phenomena, they seem to be accompanied by severe magnetic disturbances. It is, therefore, reasonable to suppose that a purposeful genesis of unusual magnetic conditions could effect a change of phase in matter, both physical and vital. If so, it would also distort the time element which is by no means an independent entity, but part-in-parcel of a particular matter-energy-time dimension such as the one we live in.

There is some interesting stuff here and Valentine seems to have inside information on the Philadelphia Experiment and had even spoken with Jessup before he died.

We learn more from some insiders in *The Philadelphia Experiment* in this conversation between Jessup and J. Manson Valentine:

[Jessup said to Dr. Valentine] The experiment had been accomplished by using naval-type magnetic generators, known as degaussers, which were "pulsed" at resonant frequencies so as to "create a tremendous magnetic field on and around a docked vessel...

The experiment is very interesting but awfully dangerous. It is too hard on the people involved. This use of magnetic resonance is tantamount to temporary obliteration in our dimension but it tends to get out of control. Actually, it is equivalent to transference of matter into another level or dimension and could represent a dimensional breakthrough if it were possible to control it.

...In practice, it concerns electric and magnetic fields as follows: An electric field created in a coil induces a magnetic field at right angles to the first; each of these fields represents one plane of space. But since there are three planes of space, there must be a third field, perhaps a gravitational one. By hooking up electromagnetic generators so as to produce a magnetic pulse, it might be possible to produce this third field through the principle of resonance.[7]

Valentine continues to comment:

The thrust of [Einstein's attempt to produce a] Unified Field Theory was a string of sixteen incredibly complex quantities (represented by an advanced type of mathematical shorthand known as tensor equations), ten combinations of which represented gravitation and the remaining six electromagnetism...

One thing that does emerge, interestingly, is the concept that a pure gravitational field can exist without an electromagnetic field, but a pure electromagnetic field cannot exist without an accompanying gravitational field.[7]

There is also this comment from an anonymous military scientist to Moore and Berlitz:

I think I heard they did some testing both along the river [the Delaware] and off the coast, especially with regard to the effects of a strong magnetic force field on radar detection apparatus. I can't tell you much else about it or about what the results were because I don't know. My guess, and I emphasize GUESS, would be that every kind of receiving equipment possible was put aboard other vessels and along the shoreline to check on what would happen on the "other side" when both radio and low- and high-frequency radar were projected through the field.

Undoubtedly observations would have also been made as to any effects that field might have had on light in the visual range. In any event, I do know that there was a great deal of work being done on total absorption as well as refraction, and this would certainly seem to tie in with such an experiment as this. [7]

Then there are these comments by an anonymous elderly scientist to Moore in the chapter "The Unexpected Key," in *The Philadelphia Experiment*:

> ...The idea of producing the necessary electromagnetic field for experimental purposes by means of the principles of resonance was... initially suggested by [physicist] Kent... I recall some computations about this in relation to a model experiment [i.e., an experiment conducted using scale models rather than real ships] which was in view at the time... It also seems likely to me that 'foiling radar' was discussed at some later point in relation to this project...
>
> The initial idea seems to have been aimed at using strong electromagnetic fields to deflect incoming projectiles, especially torpedoes, away from a ship by means of creating an intense electromagnetic field around that ship. This was later extended to include a study of the idea of producing optical invisibility by means of a similar field in the air rather than in the water...
>
> He had on one part of a sheet a radiation-wave equation, and on the left side were a series of half-finished scratches. With these he pushed over a rather detailed report on naval degaussing equipment and poked fingers at it here and there while I marked with pencil where he pointed. Then Albrecht said could I see what would be needed to get a bending of light by, oh, I think 10 percent, and would I try to complete this enough to make a small table or two concerning it...
>
> I think that the conversation at this point had turned to the principles of resonance and how the intense fields

138

which would be required for such an experiment might be achieved using this principle...

Somehow, I managed to finish a couple of small tables and a few sentences of explanation and brought all back as a memo. We went in to Albrecht, who looked it all over and said, "You did all this regarding intensities [of the field] at differing distances from the [ship's] beam, but you don't seem to pick up anything fore and aft"... All I had was the points of greatest curvature right off the ship's beam opposite this equipment...

What Albrecht wanted to do was to find out enough to verify the strength of the field and the practical probability of bending light sufficiently to get the desired "mirage" effect. God knows they had no idea what the final results would be. If they had, it would have ended there. But, of course, they didn't.

I think the prime movers at this point were the NDRC and someone like Ladenburg or von Neumann who came up with ideas and had no hesitancy in talking about them before doing any computations at all. They talked with Einstein about this and Einstein considered it and took it far enough to figure out the order of magnitude he would need on intensity, and then spoke to von Neumann about what would be the best outfit to look into it as a practical possibility. That's how we got involved in it...

I can also remember a point a little later when I suggested in a meeting of some sort that an easier way to make a ship vanish was a light air blanket, and I wondered why such a fairly complicated theoretical affair was under consideration. Albrecht took off his glasses at that point and commented that the trouble with having me at a conference was that I was good at getting them off the topic...

I do remember being at—at least one other conference—where this matter was a topic on the agenda. During this one we were trying to bring out some of the more obvious—to us—side effects that would be created by such an experiment. Among these would be a "boiling" of the water, ionization of the surrounding air, and even a "Zeemanizing" of the atoms; all of which would tend to create extremely unsettled conditions. No one at this point had ever considered the possibility of interdimensional effects or mass displacement. Scientists generally thought of such things as belonging more to science fiction than to science in the 1940's. In any event, at some point during all this, I received a strong putdown from Albrecht, who broke in with something to the effect that "Why don't you just leave these experimental people alone so they can go ahead with their project. That's what we have them for!"

One of the problems involved was that the ionization created by the field tended to cause an uneven refraction of the light. The original concepts that were brought down to us before the conference were laid out very nicely and neatly, but both Albrecht and Gleason and I warned that according to our calculations the result would not be a steady mirage effect, but rather a "moving back and forth" displacement caused by certain inherent tendencies of the AC field which would tend to create a confused area rather than a complete absence of color. "Confused" may well have been an understatement, but it seemed appropriate at the time. Immediately out beyond this confused area ought to be a shimmering, and far outside ought to be a static field. At any rate, our warning on this, which ultimately went to NDRC, was that all this ought to be taken into account and the whole thing looked at with some care. We also felt that with proper effort some of these problems could be overcome... and that a resonant frequency could probably be found that would possibly control the visual apparent internal oscillation so that the shimmering would be at a much slower rate... I don't know how far those who were working on this aspect of the problem ever got with it...

Another thing I recall strongly is that for a few weeks after the meeting in Albrecht's office we kept getting requests for tables having to do with resonant frequencies of light in optical ranges. These were frequently without

140

explanation attached, but it seems likely that there was some connection here...[7]

These curious comments by the anonymous elderly scientist to Moore seem to confirm that the Navy was experimenting with pulsed fields and such in the years during and after WWII.

In another testimonial, by an ex-military guard named Patrick

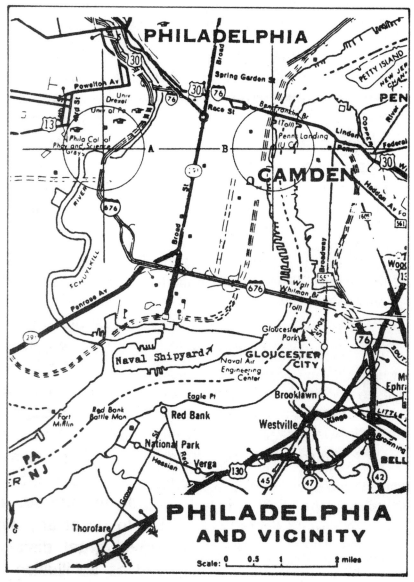

Philadelphia and vicinity map showing the location of the Naval Shipyard.

141

Macey, related a story told to him by a workmate named "Jim" in the summer of 1977:

> I was a guard for classified audiovisual material, and in late 1945 I was in a position, while on duty in Washington, to see part of a film viewed by a lot of Navy brass, pertaining to an experiment done at sea. I remember only part of the film, as my security duties did not permit me to sit and look at it like the others. I didn't know what was going on in the film, since it was without commentary. I do remember that it concerned three ships. When they rolled the film, it showed two other ships feeding some sort of energy into the central ship. I thought it was sound waves, but I didn't know, since I, naturally, wasn't in on the briefing.
>
> After a time the central ship, a destroyer, disappeared slowly into a transparent fog until all that could be seen was an imprint of that ship in the water. Then, when the field, or whatever it was, was turned off, the ship reappeared slowly out of thin fog.
>
> Apparently that was the end of the film, and I overheard some of the men in the room discussing it. Some thought that the field had been left on too long and that that had caused the problems that some of the crew members were having. [7]

This report to Moore by Patrick Macey, as told to him by a workmate named "Jim" in the summer of 1977 is interesting as it says that there was a film, without commentary, of a ship becoming invisible and reappearing. This film was being shown to Navy personnel in late 1945, shortly after the end of the war. The US Navy does not acknowledge any such film.

Moore finally summarizes what he knows about Project Rainbow and discusses some of the scientists that he has met or looked into:

> The names of several scientists have come up in connection with the revival of such a project. Two

government-employed scientists named Charlesworth and Carroll were reportedly responsible for installing the auxiliary equipment on the DE 173 [*Eldridge*] and participated in the experiment, noting the neuronal damage "due to diatheric" effect because of the "magnetic oscillation of the magnetic field."

...Victor Silverman, now living in Pennsylvania and still mindful of wartime security regulations and afraid of possible consequences, got in touch with the authors [Berlitz & Moore] through a third party when he first heard about the publication of a book about the DE 173. He speaks from personal experience: "I was on that ship at the time of the experiment."

At the outbreak of WWII Silverman enlisted in the Navy. He, along with about 40 others, was destined to become part of a special secret Naval experimental project involving a destroyer escort vessel and a process which he could identify only as "degaussing." On board the vessel, Silverman noted that there was "enough radar equipment on the ship to fill a battleship" including "an extra mast" which was "rigged out like a Christmas tree" with what appeared to be antenna-like structures.

At one point during the preparation for the experiment, Silverman remembers seeing a civilian on board and said to a shipmate: "That guy could use a haircut." To his amazement he later discovered that the man had been Albert Einstein.

Silverman was given the rating of Engineer, First Class, and, according to his account, was one of three seamen who knew where the switches were that started the operation. He also related that a special series of electrical cables had been laid from a nearby power house to the ship. When the order was given and the switches thrown, "the resulting whine was almost unbearable."

This informant, who emphatically declined to be named, confided to Berlitz that he had seen highly classified documents in the Navy files in Washington, D.C., which indicated that at least some phases of the experiment are

STILL in progress.

In addition, scientific units in private universities, some possibly funded by the government, are reported to be pursuing research in magnetic teleportation, with the attendant invisibility as part of the experiment. Some recent reports place such experimentation as having taken place at Stanford University Research Facility at Menlo Park, Palo Alto, California, and at M.I.T. in Boston. However, in the words of one informant—M. Akers, a psychologist in San Jose, California—such magnetic experiments "are frowned upon because they have detrimental effects on the researchers conducting the experiments." [7]

It is interesting to read that Albert Einstein was seen aboard the Navy ship and that the experiments are ongoing. Let us now look at the claims of Alfred Bielek, who claimed he was a survivor of the Philadelphia Experiment.

In 1984 the basic story as told to Jessup was adapted into a time travel film called *The Philadelphia Experiment*, directed by Stewart Raffill. Though only loosely based on the prior accounts of the Philadelphia Experiment, it served to dramatize the core elements of the original story. Starting in 1989, Alfred Bielek (or Al as he was commonly called) claimed to have been aboard the USS *Eldridge* during the Navy's experiment. Addressing a MUFON Conference in 1990, Bielek asserted that Raffill's film was largely consistent with the events he claimed to have witnessed in 1943. Bielek would later add details to his claims on radio talk shows, conferences, and the Internet.

Bielek was born in 1927 and died on October 10, 2011, in Guadalajara, Mexico, at the age of 84. He lived in Fort Meyers, Florida in his later years. He always claimed he had regressed to a nine-month-old baby in 1927.

His own written biography (found at bielek.com) says:

Al Bielek was born in 1927. His first memories of being Al Bielek were when he was nine months old during a family Christmas party.

The odd thing about his memory was that he fully

understood the conversation that was taking place around the piano.

As Al went through grade school, he was known as the "walking encyclopedia." Before graduating high school, he took an electronics test and was the only one to pass. The Navy needed people like him during the war years and recruited him.

Al later completed his education and took on various vocations in the field of electronics. While contracting for various

Al Bielek.

Military contractors, the people who worked with him began to reveal the truth about our involvement with Extra Terrestrials and PSI Ops (Psychic Operations) programs.

Strange things started to happen to Al soon afterwards. While in Hawaii in 1956, he had a brief encounter with, who he believes now to be Mark Hammil—the actor in Star Wars.

Soon after he was recruited into the Montauk Project.

He would work his normal job in California, and take the underground subway to Montauk Long Island to carry out his duties there.

After the time tunnel was perfected, he would simply be teleported to the underground base and returned back to his apartment.

During the 1970's, Al was the Program Director for the Psychics who manned the Montauk Chair. Since the Montauk Boys were a key program at Montauk, Al had some influence with the Montauk Boys program.

His duties were to handle the operations of the Mind Control program. He was in regular contact with Duncan Cameron and Preston Nichols. Stewart Swerdlow was one of the Montauk Boys programmers under Al Bielek.

In the 1980's when the time control programs were operational, Al participated in some of the time travel

145

experiments. Both he and Duncan traveled to Mars on several occasions. He now remembers several other trips he took with teams to a research station in 100,000 BC, other planets to get canisters filled with Light and Dark Energy, and to the year 6037.

In January 1988, after seeing the movie—"The Philadelphia Experiment," his memories started returning. Al believes his involvement with Montauk ended with that revelation. Over time and through meetings with Preston Nichols, Duncan Cameron, and others, many more memories returned. Ironically, just after his memories returned, Dr. John Von Neumann tried to get in touch with him—a promise he made to Ed Cameron, should his memories ever return.

Al made the decision to go public with the information about his involvement at Montauk and the Philadelphia experiment in 1989. He has been a prolific speaker on both radio talk shows and conferences.

He believes he has not been harmed or stopped because his time traveling experiences locked him into this timeline. Somehow, by being here today, he, among others in the program, serve to balance the effects they produced from prior time traveling experiments.

I met Al Bielek a number of times in the 1990s and early 2000s. It was typically at a conference where he and I were both speakers. While I never got the chance to thoroughly discuss things with him, I did have some talks with him and he always spoke with authority and knowledge about his amazing claims. He seemed to genuinely believe his own claims that ranged from surviving the Philadelphia Experiment, to going to Mars, to time traveling into the future to see drastic changes to our planet.

Bielek claimed that a nuclear war would devastate the globe in 2025 and that there would be flying cities by the year 2137, which he claimed to have visited. An article in the UK tabloid *The Sun* (the-sun.com/news) had an article on Bielek a short time after his death that said:

When he was just 16, he joined the US Navy to help fight the Nazis.

It was at this time in 1943, serving as a lowly naval officer aboard the USS Eldridge, that Al claimed he travelled in time.

The Eldridge was supposedly harboring the Philadelphia experiment.

On August 13, 1943, Bielek and his brother Duncan jumped off the ship to escape a strange light... and landed in the year 2137.

In a video lecture he filmed in the 1990s, Bielek said: "When Duncan and I jumped overboard, off the ship the Eldridge, in hyperspace, we didn't know what was happening and where we were going, or if we were going anywhere, except, of course, into the water.

"We expected to hit the water in the bay and swim ashore. But no water, we never hit it, we kept falling and falling for quite a period of time."

Bielek claimed the pair awoke in a hospital bed with no recollection of how they got there.

The first sign that something was wrong was that the room had a wall-mounted color TV, very unusual for 1943.

They were told by hospital staff that they had suffered severe radiation burns, not from a nuclear attack, but from the radiation found "in deep space".

Bielek claimed they were told that there were very few surviving cities around the world, and no more national boundaries or government. Eventually, the pair asked to see maps of the world and were reportedly stunned by how much it had changed in the preceding two centuries.

As Bielek explained, "much of California" was now underwater, stretching to the San Andreas faultline, and there was "not much left of Los Angeles as a functioning city." Many major US cities such as Chicago, New Orleans,

Al Bielek being interviewed.

and San Diego were gone, while rising sea levels had seen the Great Lakes become one giant body of water, and the Mississippi river widen, becoming 30 miles wide at its narrowest point.

In Europe, he said most of England had gone, while the Scottish Highlands and some of Ireland remained. Much of Europe was underwater, even parts of Switzerland. Most dramatically, he claimed that when he asked a hospital technician, he was told that the world's population had plummeted to just 300 million. He said he learned that most of the world's governments had crumbled away by the year 2025.

A nuclear world war three killed millions but, he claimed, far more were wiped out by changes to the earth.

The article also says that Bielek was walking around the hospital one day and was transported to the year 2750. In that time no militaries existed, as technology had made conflict practically impossible, and everything was now free. He said he stayed there for two years before being sent back to 2137 to pick up his brother, Duncan Cameron.

Eventually, Bielek claimed he was sent back in time, but not to the period he had come from. Instead, he appeared in 1983 on August 12 just after 2 am, apparently near Philadelphia. There, he claimed he met Dr. John Von Neumann, the Hungarian-American mathematician. This is despite the fact Dr. Von Neumann died in 1957 some quarter of a century earlier. Bielek claims that Von Neumann had not died in 1957 as is reported but continued his work with the US Navy.

Von Neumann convinced the two brothers to return to their original time and stop the Philadelphia Experiment from ever happening.

After his time in the Navy, Bielek said he was recruited by military contractors, who revealed that the US military was secretly adapting alien technology and forwarding research on psychic operations.

Soon afterwards, he was recruited to the Montauk Project, a conspiracy theory that alleges that the US government conducted

148

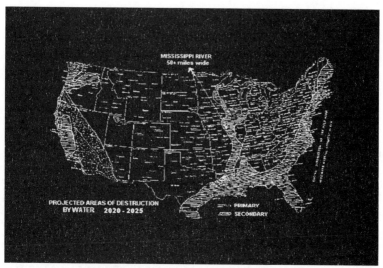

Al Bielek's map of the United States, 2020-2025.

secret projects in Montauk, New York, including psychological warfare and time travel. He claimed he was able to travel to Mars on several occasions (apparently from Montauk), as well as to a research station in the year 100,000 BC.

Bielek's time travel stories, except his working on the *Eldridge*, seem to be more like remote viewing than genuine physical time travel. His work at Montauk, according to him, involved the Montauk remote viewing chair as well as mind control. He admits that he was a subject of mind control experiments. This may be the source of many of his stories. His predictions of a global nuclear war in 2025, remain on the table, but the nor his destructions of world governments and earth changes that he said would occur between the year 2020 and 2025 have not happened. Such earth change predictions were popular in the 1990s and early 2000s but failed to occur. We can pretty much scrap Bielek's predictions as egotistic self deception.

Still, it is worth looking at some of the technical things that Bielek said early on in the 1990s. Al Bielek gave the following description of the equipment on the USS *Eldridge* to radio host Robert Barry in 1991:

> There were four RF transmitters and they were phased to produce a rotating field; they were pulsed at a 10%

duty cycle. The magnetic component of the fields was generated by four large coils set on the deck of the ship, and they were run by two large generators down in the hold of the ship—75 KVAH—they were also pulsed. The entire system was under a very special control—a very peculiar type of control—that produced a rotating field effect, which produced the interaction, which caused the time field to shift. [frequency of pulsations unknown]

[There are] FIVE DIMENSIONS [to our reality]: Our 3-D, time, and "T2," which vector was being rotated in these experiments to give time/space shifts.

All of the effects related to this are related to the PI over 2 series of frequencies, which are all WINDOWS. Although those men above deck were made insane, if not "fried," by the fields, out of this came an ELECTROMAGNETIC CANCER CURE that the Navy has "sat" on for forty years and refuses to release to the world because that would be tantamount to admitting that the Philadelphia Experiment DID happen.

The following comments on the "Zero Time Reference Generator" by Jerry Decker on Al Bielek and his claims in his January 1990 speech at MUFON were posted in the Mufon Metroplex article:

[One of] the two most outstanding things about the public lecture was the showing of a slide purported to be a "Zero Time Reference Generator," which looks strangely similar to an old Army field kitchen refrigeration unit. No technical details are usually given.

This device was purported to be the oscillator which drove the coils of the experiment. Mr. Bielek claimed that the unit shown is used to synchronize two separate signals (one for each coil). ...

Basically, a coil was wound on each half of the ship and driven by separate oscillators, synced with an adjustable phase angle to create a "Scalar-type wave." This distorted the field matrices of matter encompassed within the field for "unusual effects."

...The other interesting comment was about a Professor of Mathematics at the University of Wisconsin in the 30's. His name—Henry Levenson. He specialized in time studies and developed a time variant equation.

Mr. Levenson co-authored two books and wrote one. None of these books are available (of course) but Bielek indicates they might be available in restricted or private libraries such as Princeton.

The time comments centered around a concept involving a TIME LOCK which was "encoded" at the time of creation of all matter, living or otherwise. Thus, all matter created on the Earth, must be "clocked" to the Earth time lock. The Earth must be "clocked" to the Solar time lock and that to the Galaxial.

If your time lock became "distorted" by high intensity fields, it would create a variety of problems due to the instability of the recovery process, assuming that recovery could be possible.

Bielek claimed that the original system was powered with a 500 KW generator, later increased to 2 MegaWatt. Another experiment was supposed to have been done with 3 field coils, all synced to the same clocking system. The 3-field design created major arcing and caused a return to the 2-field design. Bielek also claims that there were 3,000 tubes used in the system. ...

The time theories espoused by Bielek seem particularly worthy of study. Much of the "Scalar" craze indicates a technology which could duplicate or surpass the original Philadelphia Experiment.

The Ghost of the Philadelphia Experiment

Recently two old books by Gray Barker have been republished by Andrew Colvin who took over Barker's Saucerian Press of Point Pleasant, West Virginia (the site of many of the Mothman occurrences). Called the New Saucerian Press, Colvin has republished, with extra material, *The Strange Case of Dr. M.K. Jessup*[16] (first published in 1963) and *The Ghost of the Philadelphia*

Experiment Returns[17] (first published in 1984). Says the back cover copy of *The Ghost of the Philadelphia Experiment Returns:*

> *The Ghost of the Philadelphia Experiment Returns* was the last book published by legendary ufologist Gray Barker. At the time, 1984, it was considered too "conspiratorial," and met with opposition in mainstream ufological circles.
>
> After compiling the book from articles in his own newsletter, Barker suddenly died, and the controversial bound edition was pulled from distribution. Luckily, New Saucerian was able to locate the prototype, and has made this wonderful work available—at last—for the general public to enjoy.
>
> In these pages, Barker deftly explores the lore of the Philadelphia Experiment, offering revelations on a variety of notorious characters, such as Carlos Allende, James Wolfe, Leon A. Seoul, Dr. Franklin Reno, Michael Ann Dunn, the Oppenheimer brothers, and William L. Moore, whose books on the Roswell UFO crash, the Bermuda Triangle, and the Philadelphia Experiment made him the highest grossing UFO author of all time. Barker also shares interesting material from researchers Morris K. Jessup, James Moseley, Dennis Pilichis, Charles Berlitz, and Anna Genzlinger, the Miami housewife who was led by Jessup's ghost to investigate his death, and who uncovered several fascinating tidbits not only about his possible murder, but also about the CIA's role in covert mind control experimentation.
>
> This special 2014 edition of *Ghost of the Philadelphia Experiment Returns* features several photos, an introduction by paranormal radio host Jeffery Pritchett, and an epilogue by the editor, Andy Colvin, whose father was stationed at the naval yard where the Philadelphia Experiment took place.[17]

The book is a compilation of material, some of it from Andrew Colvin, such as a photo of William Moore with a curious and long caption that suggests that Moore was an intelligence agent of some

152

sort. Says the caption:

This is one of the few known photographs of William L. Moore. After telling Gray Barker he had performed a study of how false, yet plausible, stories could be created from grains of truth, the reclusive Moore went on to promote some of the greatest "pro-extraterrestrial" diversion in ufological history, such as the Roswell UFO craft, the "Majestic-12" papers, and the "Aviary." Moore's early upbringing is a complete mystery, but he is known to have graduated from Thiel College (in the upper Ohio Valley, in Pennsylvania) with a degree in Russian—a common major of those in intelligence work. In the late 1960s, Moore taught high school in Mt. Perry, Ohio, where he introduced courses on ufology, and attempted to lobby the governor to make UFOs a public issue. Moore's "MJ-12" document dump featured allusions to a group of ET-believing aircraft pilots in Columbus, Ohio, called the "Quiet Birdmen"— from whom Moore seemingly modeled his "Aviary" group. After selling hundred of thousands of books to ET buffs,

William Moore.

Moore resigned from the UFO field after admitting that he had helped the Air Force manipulate Paul Bennewitz, a ufologist who "lost his mind" and eventually died in 2003.

We will discuss Moore and the MJ-12 papers in the next chapter. The book also has an interesting interview between Gray Barker and Ann Genzlinger of Miami. Genzlinger had lived in the same county as the Medical Examiner's office that had made the official investigation into Jessup's death. She was writing a book called *The Jessup Dimension* and agreed to give Barker an interview to preview her book. In the interview Genzlinger said several very interesting things:

> ...In my initial investigation, I went straight to the Medical Examiner's office, where I was given access to the autopsy file on Dr. Jessup.
> ...there was no evidence of alcohol whatsoever. But there was a complete saturation of carbon monoxide. Although this definitely establishes the cause of death, there were no other tests that could have indicated, for example, the presence of drugs—which could have been administered beforehand, by a third party.
> ...While the police were trying to revive Jessup by administering oxygen, a man who gave his name as "Dr. Harry Reed" happened to walk through the park. He examined Jessup and pronounced him dead.
> ...I have personally called and interviewed every Harry Reed in the city, and none of them have ever heard of Jessup.
> ...And this Dr. Reed was very quick to pronounce him dead. I got this from Sergeant Obenchain, who was on the case. He gave me a great deal of information. He said it was "too professional"—the way the car had been set up. He said an ordinary suicide wouldn't take the time to wet down the articles of clothing, and stuff them in the back window to make it more airtight.
> ...Also, most suicides use an ordinary garden hose. The hose used on Jessup's car was similar to one on a washing

machine—around two inches in diameter. And it was not just shoved into the car's exhaust pipe, it was *wired* on. All this had been done in broad daylight, just off a well-traveled road, at the height of the rush-hour.

...The police found no receptacles for carrying water. Jessup's clothes were not wet. Yet the rags used for stuffing windows were re-saturated with water.

...I assume you're familiar with the late Ivan Sanderson's statements. He said that Jessup got hold of a reprint of the annotated book [Varo Edition] and appended his own annotations.

...It was left at Sanderson's home. Of course, Sanderson has since died, and nobody has been able to locate it. Six months before Jessup's death, he was a guest in Sanderson's home in New York, and that's when he left the copy. He also asked Sanderson to bring three other person (we don't know just who) to Sanderson's private office. He handed them his re-annotated copy, asked them to read it, and then to lock it up for safe keeping, "in case something should happen to me." While not disclosing what the notes contained, Sanderson stated that "after having read this material, all of us developed the collective feeling of a most unpleasant nature. And this was horribly confirmed when Jessup was found dead in his car."[17]

In the reprint of *The Strange Case of Dr. M.K. Jessup* the editor Colvin has a new introduction by an early UFO researcher named Eugenia Macer-Story. Macer-Story writes in her introduction that she met Gray Barker at a very strange party at UFO investigator Jim Mosley's house. Mosley was the editor of a newsletter called *Saucer Smear* and was openly gay. Mosely has a Wikipedia page which says that he was born in 1931 and was the son of U.S. Army Major General George Van Horn Moseley, chief of the 4th Section (supplies and evacuation) of General Pershing's Wartime General Staff, and Florence Barber Moseley (née DuBois) whose family owned the Barber Steamship Lines. His parents were married in July 1930, at which time his father was already 55 years old, and James was born the following year. His childhood was spent

155

on army bases until his father's retirement in 1938.

His mother died in December 1950, leaving the nineteen-year-old James the beneficiary of a large trust fund. Moseley inherited sufficient money to be able to pursue his own interests, and he never worked a conventional career. He left Princeton University, and spent much of his time initially traveling to South America to engage in what he called "grave robbing" of pre-Columbian artifacts, then later

James Moseley circa 1980.

travelling to UFO conferences, interviewing UFO witnesses and personalities.

In 1954 Mosley cofounded the newsletter *Saucer News*. In 1968 his sold *Saucer News* to Gray Barker and then started another newsletter, *Saucer Smear*. Mosley died in Key West, Florida in 2012 at the age of 81. In his later life he admitted to partaking in a number of UFO hoaxes. Says Eugenia Macer-Story:

>...I should first comment that I did attend, as a greenhorn UFO/occult investigator and percipient, one of Jim Moseley's wild UFO-oriented parties in Fort Lee, New Jersey, in the late 1970s. I do not recall the exact year, but I was shocked by the wildness of the party, which included the screening of porn films of women and animals in a back room.
>
> Throughout the party, a conservative-appearing man in a business suit carried a Polaroid camera that he claimed was from outer space. I never returned to these party circumstances. After catching a ride back to New York City with several other UFO enthusiasts, I felt I was lucky to be alive and unscathed.
>
> However, at this gathering, I had also encountered a gentle man with a soft-spoken Southern accent: Gray Barker. At that time, Barker and I discussed the telepathic

156

and hypnotic aspects of UFO encounters, and the existence of the Tibetan tulpa, a mentally created, seemingly three-dimensional form that appears solid to the witness.[16]

Then later in the book the editor Andrew Colvin inserts a photo of Timothy Green Beckley, a UFO author and publisher known as "Mr. Creepo" with the famous porn star Hypatia Lee. In the page-long caption Colvin says:

The self-promoting Beckley also calls himself "Mr. UFO," and was instrumental in organizing the famous UFO convention in New York City in 1967. (To open the convention, Beckley invoked the protection of the "Space Brothers," while the black-turtlenecked MIB stood in the back of the room, waiting to surveil, harass, and intimidate researchers who did not believe in ETs.) Beckley has promoted some of the biggest hoaxers in ufology over his long career, but has consistently stuck to the ET line, even when writing books about how dangerous they are! For

Timothy Green Beckley with the porn actress Hypathia Lee.

over 50 years, Beckley has sold every manner of UFO, psychic, and New Age trinket through his "Bizarre Bazaar" and "Conspiracy Journal," which are modeled on old-timey huckster ads from the back of cheap comic books. Nothing at all is known about Beckley's formative years.

Timothy Green Beckley.

Following Gray Barker's death, Beckley reprinted almost all of Barker's work (including this book), along with several audiotapes, often taping over Barker's introductions in order to insert his own monologues. Recently, he has attempted to block publication of these materials by others, claiming that they are his. In his writing for porn magazines, Beckley has shown an interest in porn actresses, like Linda Lovelace and Marilyn Chambers, who claim to have been mind-controlled and forced into sex with animals. As noted in the introduction to this book, Eugenia Macer-Story saw porn films of women having sex with animals at a party hosted by Jim Mosley. The films, of course, could have been easily procured from Mosley's good pal, Beckley. In 2006, after gathering evidence that another longtime Beckley associate, Loren Coleman, had planted dubious information into hundreds of Wikipedia pages, editor Andy Colvin's website was hacked by millions of hardcore porn spam. Colvin's website had more "hits" than any other site on the West Coast, leading server owners to investigate. Their answer: the hacking was "military grade," and could not be stopped. In fact, the company, Web Eddy, was forced to shut the server completely and start their entire operation over again.[16]

Timothy Green Beckley died on May 31, 2021 in New York

City. His obituary on the Boylan Funeral Home webpages, says that he was born in New Brunswick, New Jersey, to the late Orville and Renee Beckley. It does not give any details of his early formative years as Andrew Colvin states, who says they are a mystery. Beckley does not have a Wikipedia page.

What Colvin is saying is that when he got into a fight over the rights to the Gray Barker books his website was hacked by a "military grade" hacking team that flooded his webpages from other sites. Colvin is basically saying that this had to be done by someone in the military with the capability of attacking his webpages in such a way. Colvin is suggesting that Beckley had such contacts within the military and intelligence. Why would this be? Was Mosley also an intelligence asset?

So, we are left with the uncomfortable feeling that when it comes to UFOs and the Philadelphia Experiment there are a lot of players out there who seem to have deep connections to the military and intelligence organizations that are actively watching every little thing that goes down in this field. This even includes William Moore, who we will now discuss at length in the next chapter.

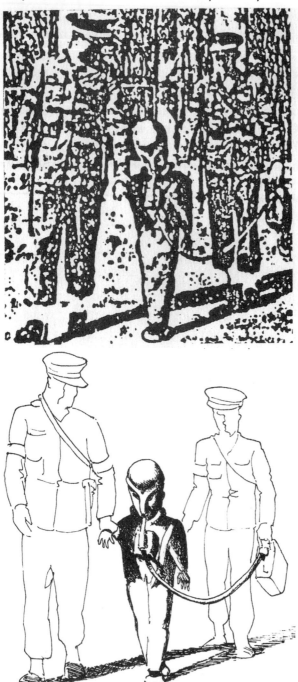

For unknown reasons Berlitz and Moore included this photocopy and drawing of a "Man from Mars" that surfaced in Wiesbaden, Germany in the late 1940s in *The Roswell Incident*. It was turned over to agent John Quinn of the FBI in 1950.

160

Chapter 8

Harry Rositzke and
The MJ-12 Documents

The wheel is turning and you can't slow down,
You can't let go and you can't hold on,
You can't go back and you can't stand still,
If the thunder don't get you then the lightning will.
—*The Wheel*, Grateful Dead

Through official secrecy and ridicule,
many citizens are led to believe the unknown
flying objects are nonsense.
—Admiral Roscoe H. Hillenkoetter
First Director of the CIA and MJ-1 of MJ-12

We have seen how Project Rainbow has led to all sorts of strange and tragic things. It has led to the death of at least one person. That person is Morris K. Jessup. Others who may have been killed because of the Philadelphia Experiment are Ian Fleming, Ivan Sanderson, James Wolf, William Sullivan and unknown others assassinated by Wernher von Braun's DISC. It is a bit of a scary scenario. We are finding out that many of our favorite UFO investigators may actually be spooks. William Moore now fits into this category.

Let us now look at a little known character apparently involved with William Moore: the leaker of the MJ-12 documents, Harry Rositzke. Who was Harry Rositzke? Harry Rositzke (who died in 2002 at the age of 91) was an American spymaster whose career included researching the origins of the English language and probing the inner workings of Nazi Germany and, later, the Soviet Union. For 25 years, he ran CIA covert operations against the Soviet Union from several overseas posts as well as Washington.

It has been said by UFO investigator and author Bill Moore

and others that Rositzke was "Falcon" and the head of the Aviary, a group of individuals, mainly civilians, who had worked for the CIA. Moore has said that Rositzke was the one who leaked the documents to him and Jamie Shandera, warning them that there was truth and disinformation in the documents.

Harry Rositzke.

Another person we will look at in this chapter is Roscoe H. Hillenkoetter, a US Navy Admiral and the first head of the CIA and the top member of the MJ-12 list. He was famous in the 1940s as an Admiral in the US Navy during WWII but is not well known today.

Harry Rositzke, the Aviary and the Release of the MJ-12 Documents

Harry Rositzke was a long-time CIA operative and had worked with the Office of Strategic Services (OSS, the precursor of the CIA) during WWII. He was born in Brooklyn in 1911 and lived on his farm in Warrenton, Virginia until he died on November 4, 2002.

He was a graduate of Union College in Schenectady, New York, where he earned his A.B. in 1931, and of Harvard University, where he received his Ph.D. in Germanic philology in 1935; he also studied phonetics at the University of Hamburg from 1935 to 1936. The first six years of his career were spent teaching English at Harvard University, the University of Omaha, and the University of Rochester. When the United States entered World War II in 1942, Rositzke joined the army and became a major. After the war he was hired by the OSS to monitor Soviet intelligence activities. He remained with the CIA until 1970, working in posts in Munich and New Delhi and retiring as chief of the international Communism unit. Said a *Washington Post* obituary on November 7, 2002:

162

Harry Rositzke, 91, a farmer, author, teacher, scholar and spy who for 25 years ran Central Intelligence Agency covert operations against the Soviet Union from Munich, New Delhi, New York and Washington, died of pneumonia Nov. 4 at Fauquier Hospital in Warrenton. Mr. Rositzke raised beef cattle in Middleburg, wrote books about the CIA and the KGB, taught at Harvard University, researched such arcana as Anglo-Saxon grammar and vowel duration in High German, and, during the Cold War, directed the parachuting of espionage agents into the Ukraine region of the Soviet Union.

His books include *The CIA's Secret Operations* (1977) and *The KGB: The Eyes of Russia* (1981), but he also wrote "The C-Text of the Old English Chronicles," which is considered a classic of Anglo-Saxon research. He did his doctoral dissertation at Harvard on "The Speech of Kent Before the Norman Conquest."

In 1981, *The Washington Post*'s Henry Allen wrote of Mr. Rositzke, "He should be George Smiley, the John LeCarre spy novel hero—a little out of the mold, a scholar." Responding to that appraisal, Mr. Rositzke said: "I was just reading *Smiley's People*. The point is, academic training leads you to look at the facts, to weigh the facts. But Smiley couldn't exist in any real environment."

Mr. Rositzke was a veteran of World War II duty with the Office of Strategic Services, the predecessor in espionage to the CIA. He volunteered in 1946 to monitor the intelligence operations of the Soviet Union, a major wartime ally against Nazi Germany. In the OSS, he had been chief of military intelligence in London and Paris, and later chief of the steering division in Germany, where he operated out of a former sparkling-wine factory near Wiesbaden.

Arthur M. Schlesinger Jr., who became an aide to President John F. Kennedy and a presidential scholar, was one of Mr. Rositzke's OSS colleagues. It came to him as no surprise that Mr. Rositzke opted for a career in intelligence after the war. "War had made him a professional. Peace evidently offered him a scope for analysis and action on questions more urgent than Anglo-Saxon grammar, his previous specialty," Schlesinger wrote in a preface to Mr. Rositzke's 1977 book on the CIA.

Mr. Rositzke, a native of Brooklyn, N.Y., graduated from Union College and received a doctorate in Germanic philology at Harvard. In 1935 and 1936, he studied experimental phonetics on a fellowship at the University of Hamburg, where he had an opportunity to witness the Nazis' consolidation of power in Germany. He would later describe this experience as frightening. From 1936 to 1942, he taught at Harvard, the University of Omaha and the University of Rochester.

As a specialist on Soviet intelligence after the war, he moved initially into quarters in Washington where, by his own description: "The walls were pockmarked with holes and the ceiling smudged with stains from the rain and snow that leaked through the fragile roof. It had no carpet. It was furnished with one antique desk."

He ran agents in the Soviet Union and Eastern Europe from 1949 to 1954 and was based in Munich for the last two years of this assignment. "We were sending people into the Ukraine—people forget that there was an active resistance movement there. ... We'd fly them in and parachute them from C-47s. We never lost a plane. We were pleased to see how inefficient the antiaircraft forces were." The East German government in those years gave Mr. Rositzke one of the longer entries in its published directory of CIA agents operating in the region.

As the CIA station chief in New Delhi from 1957 to 1962, Mr. Rositzke's espionage targets were Soviets, Chinese and Tibetans. He lunched monthly with his resident counterpart from the GRU, the Soviet military intelligence arm, and he also developed a working relationship with John Kenneth Galbraith, Kennedy's ambassador to India, who was deeply suspicious of the CIA.

Later in the 1960s, he worked on the recruitment by the CIA of Soviet diplomats in Washington and New York and began to focus on terrorism and wars of national liberation. He retired from the CIA in 1970 as chief of the international communism unit.

In 1955, Mr. Rositzke purchased a 350-acre farm in Middleburg, where in retirement he raised beef cattle, mowed his fields, drove his tractor and wrote about the craft of intelligence.

Espionage, he would argue, had a useful role in the maintenance of political order. "Spies in the right places can induce a feeling of security by negative reporting or guarantee no strategic surprises by positive reporting. Their value in reducing the paranoid tendencies of the Soviet Union should not be underestimated."

Writing about the KGB, Mr. Rositzke observed: "The clandestine mentality is rooted in a conspiratorial view of the world ...[that] someone out there is plotting against me.... Since the world is a threatening place, only secret counter-action can guarantee survival." In a 1981 *Washington Post* review, David Wise wrote that this argument "could also provide the rationale for the CIA's own covert operations around the globe."

The MJ-12 Files and Rositzke

We can see that Harry Rositzke is quite an interesting person and someone who knew a lot of people in intelligence—not just in the CIA but around the world. In the 1940s, 50s and 60s he was in Europe, India, Tibet and no doubt many other places. He would be an ideal person to send the MJ-12 documents to Jamie Shandera and Moore if the CIA needed such a person. Rositzke must have known a great deal about the SS and its continued activities after the war.

John Greenwald Jr.'s Black Vault Internet site of Freedom of Information documents does have a few documents for Rositzke. They do not tell us much but one document is a June 29, 1961 FBI document that accuses Rositzke of being a Communist. The document is apparently from a relative of his wife's who complains that she stopped being a good Catholic after getting married to Rositzke and that Rositzke had espoused some positive views towards the Soviet Union.

What can be noted here is that this was what the 1950s and the early 1960s were about: if your neighbor is a Commie then you need to turn them in. This probably meant writing a letter to the FBI alerting them to your suspicions. In this case, the perpetrator is a veteran CIA agent. Still, the FBI, in its paranoia about the CIA and other intelligence groups in the USA, thought it was proper to consider this accusation against Rositzke and keep it in their files. But otherwise, Rositzke managed to keep his head down during these difficult times and seems to have been in the good graces of the CIA during all these years. He must have known a great deal about the Operation Paperclip affairs, including the secret SS submarine base in the Canary Islands.

Indeed, although Rositzke wrote a number of memorable books he has no Wikipedia page and is only mentioned casually in discussions on the Internet and elsewhere. Like many characters from the strange UFO activities beginning in WWII, Rositzke was aware of much of what was going on, but his masters were the CIA. Naval Intelligence and the FBI were not his employers and they apparently distrusted the CIA and its collaboration with former SS officers and Operation Paperclip Germans who were being brought to the US.

As far as MJ-12 is concerned, it seems that it was an operation by some element of the government to get information into the UFO community. Moore told radio host Greg Bishop that Rositzke had been called out of retirement to conduct the operation of leaking the MJ-12 documents.

Moore told Bishop that the plan to release the MJ-12 documents had been hatched before Rositzke was brought in. Therefore the idea for the operation and the initial direction to Rositzke was coming from a higher level—Falcon had a boss.

It is speculated that the person who brought Rositzke into the UFO game was Richard Helms who outranked Rositzke, being a former Director of the CIA. There are several stories about Helms and his eagerness to share UFO material with certain UFO researchers. Moore also suggested to Greg Bishop in his radio interview that his colleague Jamie Shandera had previous contacts with the CIA before he received the documents. Moore also

166

admitted that he had contacts with the CIA when he was at university.

We should remember that the document drop of the MJ-12 documents to Shandera was in 1984. This is about five years before the document drop of German Haunebu, Vril, and other material was made to Ralph Ettl while he was living

Richard Helms in 1974.

in London. There is a gnawing similarity here between the two document dumps; they are both coming from intelligence sources and both have to do with UFOs and their activity starting in the 1940s.

The MJ-12 documents, brief as they are, only tell us two things: that there was a group that studied UFOs—and here are their names—and that there were several flying saucer crashes at Roswell. Moore had recently coauthored the book *The Roswell Incident*[29] with Charles Berlitz on Roswell. What was the disinformation—that MJ-12 was real or that several flying saucers with aliens crashed at Roswell? It seems that MJ-12 was a real group, a cabal if you will. MJ-12 probably does not exist today and it may have been dissolved in the 1950s, shortly after the Korean War. Let us look at MJ-12 and the documents provided to Shandera and Moore.

1947: The Creation of Majestic-12

MJ-12 was a group of twelve men who were operating a committee to investigate "alien spacecraft" per an executive order from President Truman on September 24, 1947. This secret committee of scientists, military leaders, and government officials was to be called by a code name: Majestic-12 (or MJ-12).

During the early 1980s, several books were published concerning a cover-up of the Roswell UFO incident of July 1947; the authors speculated some secretive upper tier of the United States government or military was responsible. Then, in 1984 UFO researcher Jaime Shandera received an envelope

167

containing film that, when developed, showed images of eight pages of documents that appeared to be briefing papers describing "Operation Majestic-12." The documents purported to reveal a secret committee of 12 scientists and military officers that existed in 1947. This committee was to look into how the recovered alien technology could be exploited, and how the US should engage with extraterrestrial life in the future.

Shandera, with his colleagues Bill Moore and Stanton T. Friedman, said they later received a series of anonymous messages that led them to find another document that has been called the "Cutler/Twining memo." This memo purports to be written by President Eisenhower's assistant Robert Cutler to General Nathan F. Twining. It adds authenticity to the MJ-12 documents because it contains a reference to Majestic-12. The memo is dated July 14, 1954 and refers to the scheduling of an MJ-12 briefing. General Twining is MJ-4, according to page two of the documents. The government says that the documents are all fakes.

The following individuals were described in the documents as "designated members" of Majestic-12 in this order:

Roscoe H. Hillenkoetter (MJ-1)
Vannevar Bush (MJ-2)
James Forrestal (MJ-3)
Nathan F. Twining (MJ-4)
Hoyt Vandenberg (MJ-5)
Detlev Bronk (MJ-6)
Jerome Clarke Hunsaker (MJ-7)
Gordon Gray (MJ-8)
Donald H. Menzel (MJ-9)
Sidney Souers (MJ-10)
Robert M. Montague (MJ-11)
Lloyd Berkner (MJ-12)

The document then noted that General Walter B. Smith replaced James Forrestal on August 1, 1950 because of Forrestal's mysterious death that we will now discuss.

The Death of James Forrestal

Forrestal was the last Cabinet-level United States Secretary of the Navy and the first United States Secretary of Defense. He was nominated to be Undersecretary of the Navy by President Franklin D. Roosevelt in 1940, and he led the national effort for industrial mobilization for the war effort during World War II. He was named Secretary of the Navy in May 1944, and was the first Secretary of the newly created Defense Department in 1947 by Roosevelt's successor Harry S. Truman. Forrestal signed the order for Operation Highjump and the invasion of Antarctica. Forrestal was also supposedly a member of the Top Secret government control group known as MJ-12. He died under mysterious circumstances in 1949.

Forrestal was a big supporter of naval battle groups that were centered around aircraft carriers. As Secretary of Defense, it is said that he was often at odds with President Truman over national policy. As the Defense Department drew down after the war Forrestal fought for all the money he could get to keep the American Army from shrinking.

In 1948 Thomas Dewey was widely expected to win the Presidential election, and it was later revealed that Forrestal had met and negotiated for a cabinet position with Dewey, who was Truman's opponent. But Truman won the election and Forrestal was forced to resign as Secretary of Defense by the president in 1949. Shortly after his resignation he underwent medical care for depression, and died after falling from a sixteenth floor window of the hospital where he was being treated.

Forrestal was born in Matteawan, New York, the son of Irish immigrants. His mother raised him as a devout Roman Catholic. He was an amateur boxer. After graduating from high school at the age of 16, in 1908, he spent the next three years working for a trio of newspapers: the *Matteawan Evening Journal,* the *Mount Vernon Argus* and the *Poughkeepsie News*

James Forrestal.

169

Press.

Forrestal entered Dartmouth College in 1911, but transferred to Princeton University in his sophomore year. He served as an editor for *The Daily Princetonian*. In 1926 Forrestal married Josephine Stovall a *Vogue* writer.

When the United States entered World War I, he enlisted in the Navy and eventually became a Naval Aviator, training with the Royal Flying Corps in Canada. During the final year of the war, Forrestal spent much of his time in Washington, DC, at the office of Naval Operations while completing his flight training, and eventually reached the rank of Lieutenant.

After the war, Forrestal worked in finance and made his fortune on Wall Street. He also acted as a publicist for the Democratic Party committee in Dutchess County, New York helping politicians from the area win elections at both the state and national levels. One of those individuals aided by Forrestal's political work was a neighbor and fellow Democrat, Franklin D. Roosevelt.

When Franklin D. Roosevelt became president he appointed Forrestal as a special administrative assistant on June 22, 1940. Six weeks later, he nominated him for the newly established position, Undersecretary of the Navy. In his nearly four years as undersecretary, Forrestal proved highly effective at mobilizing domestic industrial production for the war effort.

He became Secretary of the Navy on May 19, 1944, after his immediate superior Secretary Frank Knox died from a heart attack. Forrestal led the Navy through the closing year of the war and the painful early years of demobilization that followed. As Secretary, Forrestal introduced a policy of racial integration in the Navy.

Forrestal traveled to combat zones to see naval forces in action. He was present at the Battle of Iwo Jima in 1945, and claimed the historic flag hoisted over Japanese soil as a souvenir. A second, larger flag was run up in its place, and this second flag-raising was the moment captured by Associated Press photographer Joe Rosenthal in his famous photograph.

Forrestal, along with Secretary of War Henry Stimson and Undersecretary of State Joseph Grew, in the early months of 1945, strongly advocated a softer policy toward Japan that would permit a negotiated armistice, a 'face-saving' surrender. Forrestal's primary concern was not the resurgence of a militarized Japan, but rather "the menace of Russian Communism and its attraction for decimated, destabilized societies in Europe and Asia." Forrestal was for keeping the Soviet Union out of the war with Japan so the nation could not overly exert its influence in the area.

Strongly anti-Soviet and anti-Communist, after the war, Forrestal urged President Truman to take a hard line with the Soviets over Poland. He also strongly influenced the new Wisconsin Senator, Joseph McCarthy. Upon McCarthy's arrival in Washington in December of 1946, Forrestal invited him to lunch and shared with him his concerns about Communist infiltration of the US, including the government.

In early 1946 Forrestal authorized the invasion of Antarctica, and Operation Highjump came into effect. Discussed at length in my book *Antarctica and the Secret Space Program*, Operation Highjump was a United States Navy operation organized by Rear Admiral Richard E. Byrd Jr. "Task Force 68" included 4,700 men, 13 ships, and 33 aircraft. Operation Highjump's primary mission was said to be the establishment of the Antarctic research base Little America IV, but stories persist that Operation Highjump was a military invasion of Antarctica that has never been fully explained. This operation was apparently an attempt to subdue the German U-boat base in Neuschwabenland that continued to operate after the war.

At the time MJ-12 was being formed—1947—President Harry S. Truman appointed Forrestal the first United States Secretary of Defense. During this time Forrestal continued to advocate for complete racial integration of the military services, a policy that was eventually implemented in 1949.

Forrestal had argued against the partition of Palestine to create the new nation of Israel on the grounds it would infuriate Arab countries who supplied oil needed for the US economy and national defense.

Forrestal's stance soon earned him the active enmity of some congressmen and supporters of Israel. Forrestal was also an early target of the muckraking columnist and broadcaster Drew Pearson, an opponent of foreign policies hostile to the Soviet Union, who began to regularly call for Forrestal's removal after

James Forrestal meets with President Harry Truman.

President Truman named him Secretary of Defense. Pearson, who was known for dirty tactics, once told his own protégé, Jack Anderson, that he believed Forrestal was "the most dangerous man in America" and claimed that if he was not removed from office, he would "cause another world war."

His 18 months as Secretary of Defense had been a difficult time for the US military establishment: Communist governments came to power in Czechoslovakia and China; the Soviets had imposed a blockade on West Berlin which prompted the US-Berlin Airlift to supply the city; on May 14, 1948 the State of Israel was declared and the Arab–Israeli War followed. Also, negotiations were going on for the formation of NATO.

In 1948 the Governor of New York, Thomas E. Dewey, was expected to win the presidential election. Forrestal had a private meeting with Dewey and it was agreed he would continue as Secretary of Defense under a Dewey administration. Then, just weeks before the election, Forrestal's nemesis, the radio personality and journalist Drew Pearson, published an exposé of the meetings between Dewey and Forrestal.

Truman was angered by Forrestal's maneuvering behind his

172

back and his resistance to the military draw down that Truman had been ordering. The President abruptly asked the Secretary of Defense to step down. By March 31, 1949, Forrestal was replaced by Louis A. Johnson who was a firm supporter of Truman's policies. Around this time, Forrestal had a nervous breakdown and believed that he was being followed and that his phone calls were being monitored.

On the day of Forrestal's resignation from office, March 28, he was reported to have gone into a strange daze and was flown on a Navy airplane to the estate of Undersecretary of State Robert A. Lovett in Hobe Sound, Florida, where Forrestal's wife, Josephine, was vacationing. Dr. William C. Menninger of the Menninger Clinic in Kansas was consulted about Forrestal's strange condition and he diagnosed "severe depression" of the type "seen in operational fatigue during the war."

It was suggested that Forrestal should go to the Menninger Clinic in Kansas, but Forrestal's wife, along with friend Ferdinand Eberstadt and Navy psychiatrist Captain Dr. George N. Raines, decided to send the former Secretary of Defense to the National Naval Medical Center (NNMC) in Bethesda, Maryland.

Forrestal was checked into NNMC five days later. The decision to house him on the 16th floor instead of the first floor was justified because of a need to keep Forrestal's mental health a secret. Forrestal's condition was officially announced as "nervous and physical exhaustion"; his lead doctor, Captain Raines, diagnosed his condition as "depression" or "reactive depression."

Captain Raines reportedly gave Forrestal the following drugs:

1st week: narcosis with sodium amytal.
2nd–5th weeks: a regimen of insulin sub-shock combined with psycho-therapeutic interviews. According to Dr. Raines, the patient overreacted to the insulin much as he had to the amytal and this would occasionally throw him into a confused state with a great deal of agitation and confusion.
4th week: insulin administered only in stimulating doses; 10 units of insulin four times a day, morning, noon, afternoon and evening.

According to Dr. Raines, "We considered electroshock but thought it better to postpone it for another 90 days. In reactive depression if electroshock is used early and the patient is returned to the same situation from which he came there is grave danger of suicide in the immediate period after they return... so strangely enough we left out electroshock to avoid what actually happened anyhow."

What was happening to Forrestal? His life was falling apart. He was fired from his extremely important job. He had delusions and acted in a daze. He thought people were following him and listening to his phone calls. Was he being drugged? Were people actually following him and listening to his phone calls? Had he just lost his marbles, as they say? All of these are possible. One other thing to consider—was Forrestal (MJ-3) still part of the secret group designated MJ-12? It would seem so.

Forestal at the National Naval Medical Center

Since entering the NNMC Forrestal gained 12 pounds and seemed to be on the road to recovery. However, in the early morning hours of May 22, 1949, his body, clad only in the bottom half of a pair of pajamas, was found on a third-floor roof below the sixteenth-floor kitchen across the hall from his room.

The official Navy review board, which completed hearings on May 31, waited until October 11, 1949 to release only a brief summary of its findings. The announcement, as reported on page 15 of the October 12 *New York Times*, stated only that Forrestal had died from his fall from the window. It did not say what might have caused the fall, nor did it make any mention of a bathrobe sash cord that had been reported by the coroner as being tied around his neck.

A guard had been assigned to watch Forrestal at all times, sitting in the room with the man who held some of the most important secrets in the government. They were on the sixteenth floor of NNMC in a room where the windows had been altered so that they could not open. The main guard assigned to Forrestal was US Navy corpsman Edward Prise. Prise and Forrestal had become good friends and Forrestal even said shortly before his death that he wanted Prise to be his driver when he was eventually released from the hospital.

On the night of Forrestal's death, Prise had been sitting in the room until it became fairly late and Prise's shift ultimately ended. Forrestal told Prise that he didn't need a sleeping pill for that night and was going to stay up and read for a while. Prise was replaced by another Navy corpsman named Robert Wayne Harrison Jr. At some point in his shift Harrison left the room to take his five-minute break. When he returned to the room Harrison said that he was shocked to find Forrestal gone. Racing to find him, Harrison noticed that the cord from Forrestal's robe was tied to a radiator near the open window of the small kitchen across the hall. Looking down from the window he could see Forrestal's dead body 13 floors below him. Had Forrestal tried to hang himself with the cord from his robe as he climbed out of the window—but instead fell to his death—or had he been pushed?

Why had Harrison been able to leave his post, apparently against orders that Forrestal was not to be left alone? Forrestal was due to be released soon said his brother Henry Forrestal shortly afterwards. Henry Forrestal also believed that his brother had been murdered, probably by the Navy.

Incredibly, a full report was not released by the Department of the Navy until April 2004 in response to a Freedom of Information Act request by researcher David Martin. The report said:

> After full and mature deliberation, the board finds as follows:
>
> FINDING OF FACTS
> That the body found on the ledge outside of room three eighty-four of building one of the National Naval Medical Center at one-fifty a.m. and pronounced dead at one fifty-five a.m., Sunday, May 22, 1949, was identified as that of the late James V. Forrestal, a patient on the Neuropsychiatric Service of the U. S. Naval Hospital, National Naval Medical Center, Bethesda, Maryland.
> That the late James V. Forrestal died on or about May 22, 1949, at the National Naval Medical Center, Bethesda,

175

Maryland, as a result of injuries, multiple, extreme, received incident to a fall from a high point in the tower, building one, National Naval Medical Center, Bethesda, Maryland.

That the behavior of the deceased during the period of his stay in the hospital preceding his death was indicative of a mental depression.

That the treatment and precautions in the conduct of the case were in agreement with accepted psychiatric practice and commensurate with the evident status of the patient at all times.

That the death was not caused in any manner by the intent, fault, negligence or inefficiency of any person or persons in the naval service or connected therewith.

With this last statement the US Navy is essentially saying that it investigated itself and found that it was not to blame. Naval Intelligence and other naval groups were not somehow involved. Unfortunately it is difficult to take this statement at face value. One of the difficulties with intelligence services is that it is hard to get accurate statements from them in general, let alone when they undertake investigations in which they themselves are the accused perpetrators.

According to Nick Redfern in his book *Assassinations*,[53] only a few days before Forrestal was sent to Florida and eventually to the hospital, he was visited by his friend Ferdinand Eberstadt. When Eberstadt called to say he wanted to stop by, Forrestal said in a strange voice, "For your own sake, I advise you not to." Eberstadt was a lawyer and a banker who said he was stunned to find all of the curtains shut at Forrestal's home.

In hushed tones Forrestal told him that there were listening devices all over the house. He also said that his life was in danger and that sinister forces were watching his every move. Forrestal then opened one of the blinds and pointed to two men who were standing on a street corner and assured Eberstadt that they were part of the plot and were watching him.

Just then the doorbell rang which threw Forrestal into a panic. A staff member in the house answered the door and spoke with the person there. The staff member then came into the room and told Forrestal that the visitor wanted to speak with him as he was trying to gather support to become the postmaster in his hometown. Could he come in and speak with Forrestal, who might have some advice for him? Forrestal refused to speak with the man and then Eberstadt said that the two watched through an open blind as the man walked directly to the two men on the corner and began speaking with them. This was more evidence of a conspiracy against him, he told Eberstadt grimly.

Eberstadt was never able to figure out if it was just a coincidence that the man talked with the strangers on the corner or not. Perhaps Forrestal was being watched and harassed. Redfern, in his book, suggests that Forrestal was visited by one of the Men in Black who began cropping up around this time. Redfern also thinks that Forrestal was paranoid for good reason and ultimately silenced before he could be let out of the hospital. He was a man who knew too much.

In his 2019 book *The Assassination of James Forrestal*,[38] author David Martin maintains that Forrestal was murdered while at the NNMC. Throughout his book he says that Forrestal was murdered by pro-Israel activists that operated within the government and the US Navy. In light of Forrestal's well-known opposition to the partition of Palestine and the creation of Israel it is easy to see why this may have been a motive for Forrestal's "assassination."

However, his character had already been assassinated by the pro-Israel and pro-Soviet radio personalities, Drew Pearson and Walter Winchell, and he had been removed from power. Was the pro-Israel lobby, powerful as they were, able to have Forrestal harassed and then assassinated at a Navy hospital? It would seem that Forrestal may have been assassinated, but it is more likely because of his instability and his role in MJ-12 rather than his opposition to Israel.

What happened to James Forrestal? What did he know as a member of MJ-12? Did he know things about Operation Highjump and Antarctica that were too top secret to be discussed with other politicians—something that Forrestal was known to do? Forrestal reportedly saw himself as a "behind the scenes" sort of person, but his importance and stature may have been just too large for even him to control.

Once Forrestal was forced out by President Truman did he become a target for mind control and psychological harassment, perhaps with drugs or other methods? Once a person was part of such an insider group as MJ-12, was it really possible to go back to normal civilian life? I think probably not. A combination of drugs and genuine harassment and surveillance would be enough to drive most men to some form of paranoid behavior, much like Forrestal was displaying.

The MJ-12 Documents

Aside from Forrestal's strange death, in the late 1940s the US was giddy with excitement that World War II was finally over with the surrender of Japan in the summer of 1945. But, while the rest of the world could go on with its business, within two years the top intelligence officials in the US government, guided by Naval Intelligence and directed by Admiral Roscoe H. Hillenkoetter, would face a new challenge—a puzzling onslaught of Unidentified Flying Objects. Sometimes flying in a V-formation, these craft were seen flying around various parts of the world, including the United States. Had World War II not ended two years earlier?

Because of this new challenge the secret group called MJ-12 was apparently formed. Also taking place during this time period was the infamous Project Paperclip, the recruitment of former Nazis and German officers, some of them scientists, like Wernher von Braun, to come and work in the United States.

When we look at the puzzling MJ-12 documents we seem to get a gnawing feeling that there is something wrong. Operation Majestic-12 appears to be a very real group of twelve senior intelligence officers and scientists within the US government, and all of the people listed in the papers existed.

Some of the names will be instantly familiar to those who study history, while others are more obscure. It is beyond the scope of this book to discuss all of the people on this original MJ-12 list, but each is a fascinating person.

Let me just say that Admiral Roscoe Henry Hillenkoetter (May 8, 1897—June 18, 1982) was clearly the top intelligence officer in the US government immediately after the war. He was the third

Roscoe H. Hillenkoetter in 1957.

director of the post–World War II United States Central Intelligence Group (CIG), the third Director of Central Intelligence (DCI), and the first director of the Central Intelligence Agency created by the National Security Act of 1947. He served as DCI and director of the CIG and the CIA from May 1, 1947 to October 7, 1950. He was also the head of MJ-12, according to the document.

Also according to the document, General Walter B. Smith replaced James Forrestal on the MJ-12 panel on Aug. 1, 1950 after Forrestal's death. This makes General Smith a thirteenth member of the original group. A fourteenth name is given to us in the document, that of President Dwight Eisenhower. The document also states that President Truman created Operation MJ-12 by an executive order on September 24, 1947.

But what of the other contents of this document dump? The third page mentions the Kenneth Arnold sighting of boomerang-shaped objects over the Cascade Mountains of Washington State on June 24, 1947. While Arnold described them as boomerang-type aircraft, very similar to the German Horton IV flying wing, the MJ-12 document, page three, calls this and the many other

The *Roswell Daily Record* front page on July 8, 1947.

sightings "disks." And, indeed, there were many sightings of disks, rather than boomerang-type shapes.

The next paragraph discusses the July 7, 1947 crash of a "disk" near the Air Force base in Roswell, New Mexico, an area that is heavily militarized. The next paragraph mentions the wreckage of a craft and the partially destroyed remains of "four small human-like beings" that had "ejected" from the craft before it exploded and crashed. It also said that the bodies of the four small beings had lain in the desert for approximately a week before they were found and recovered. Yes, one week!

The next page of the MJ-12 documents begins with an assessment that the small saucer craft are scout craft because they have no amenities in them, such as food or a toilet. The document says that Dr. Bronk suggested they call the occupants of the craft EBEs for "Extraterrestrial Biological Entities." The next paragraph says that the occupants of the Roswell craft are thought by Dr. Menzel to be from another solar system.

The third paragraph says that there are symbols in the craft that appear to be writing, and that there is no electrical wiring or other recognizable electronic components. The last sentence says that it was "assumed the propulsion unit was completely destroyed by the explosion which caused the crash."

180

The next page, page five, starts by saying that there is a need for more information and mentions other secret Air Force projects: Project Sign and Project Grudge. It then mentions another operation that is coordinating these operations. This operation is called Blue Book.

The second paragraph mentions that another UFO crashed on December 6, 1950 on the Mexican side of the Texas-Mexico border near a town called El Indio. The craft was mostly destroyed but what little that was recovered was sent to Sandia Air Force base just east of Albuquerque, New Mexico.

The third and final paragraph talks about how we do not know the ultimate intentions of these "visitors" and that we need to keep studying them as well as keep the public calm and not cause a panic. It then mentions a mysterious contingency plan called MJ-1949-04P/78. We never learn what it is, except that it is Top Secret and is attachment "G."

The next pages of the documents are a list of the various attachments, and the final pages are the original memorandum addressed to Forrestal authorizing the creation of MJ-12—spelled out in the document—signed by President Truman and dated September 24, 1947. We also have the supplementary document called the Twining Memo which mentions a briefing by MJ-12.

While the government has denied that the MJ-12 documents are valid, many researchers in the UFO community felt at the time that they were authentic. Curiously, the skeptics were divided in their opinions on who had created the fake documents. Some pointed to UFO researchers Jamie Shandera and Bill Moore as the creators of the supposed hoax. But, other skeptics felt that the forgery had occurred—get this—within the US government itself in order to plant disinformation with Shandera and his colleagues. In this theory, the CIA or Air Force Intelligence, or some other intelligence agency, had created these fake documents about MJ-12 in order to spread disinformation.

This "disinformation" was handed anonymously to several of the top UFO researchers in the 1980s. But why would there be disinformation like this? In most instances of disinformation, there is some real information. What was trying to be hidden with fake documents about MJ-12? What was the real information? Was it to fool gullible UFO researchers that there was once a secret MJ-12 group within the government and intelligence agencies that never existed? Would four dead and decomposing extraterrestrials really be lying out in the desert for a week after a UFO crash only miles from a major Air Force base?

There is nothing really classified about the discussion of the Kenneth Arnold sighting or the Roswell crash. All of this was largely "public knowledge" and there is not much real meat in this Top Secret briefing, except for the creation of MJ-12 itself. This group and its members is the real story, whatever their activities may have been.

Is it possible that what was really discussed by MJ-12 were the results of the invasion of Antarctica the year before? With all of the unusual events happening at the end of World War II, including UFO sightings and a large number of missing submarines, it would seem that the most pressing issues were still-existing Nazi bases in various parts of the world, as well as the rise of the Soviet Union and other events in China, Korea and Vietnam.

It would seem that the MJ-12 documents were hiding a larger threat than small human-like EBEs. And if the claims about extraterrestrials in the MJ-12 documents are genuine, did the group conclude that they were coming from Antarctica? Operation Highjump, an invasion of Antarctica, ended in late February of 1947 and the MJ-12 group was formed in September later that year.

If those participating in Operation Highjump witnessed flying saucers in Antarctica, as has been reported, then perhaps MJ-12 was formed as a think tank to be the "spin doctors" on the strange events happening around the globe and the flying saucer craze that swept through Europe, Asia, the Americas and the world. Where were these craft coming from—outer space? Or, were they coming from Antarctica? Was this why Forrestal was drugged, hospitalized, and murdered? He was too vocal in his discussions with other people, including Senator McCarthy—famously so. Can this be the reason he ultimately ended up in a Navy hospital being given all sorts of mind control drugs?

Furthermore, Forrestal's anti-Soviet stance was causing trouble with the prevailing idea of fostering some cooperation with the

Russians while maintaining an adversarial position on the world stage. Ultimately the Cold War was one of cooperation, as were the space initiatives—the "space race"—culminating with the International Space Station and the dissolution of the Communist Soviet Union in 1989. How can you have a war with someone and share a space station with that entity at the same time? Ask NASA.

Something was happening in Antarctica and it didn't have to do with the Russians. They had plenty of frozen territory in the north to deal with. Operation Highjump had returned from Antarctica and had shaken the Navy with what they had discovered. James Forrestal knew too much about this activity and was considered a loose cannon on the deck. He had to be eliminated by first sending him to the Navy loony bin and then having him commit suicide. MJ-12, however, would continue on without him.

> Once they have you asking the wrong questions,
> they don't have to worry about the answers.
> —Thomas Pynchon, *Gravity's Rainbow*

A Look at the Members of MJ-12

Let us have a look at the members of MJ-12 other than President Truman, Forrestal and Admiral Hillenkoetter. One that stands out is the famous UFO debunker astronomer Donald Menzel.

Donald Howard Mezel (April 11, 1901—December 14, 1976) was an astronomer, President of the American Astronomical Society from 1954 to 1956, and a renowned, though incompetent, UFO debunker. In his three anti-UFO books, he argued that all unidentified objects were simply misidentified natural phenomena.

For instance, in his 1963 book *The World of Flying Saucers* (co-written by Lyle Boyd) Menzel maintains that air inversions are the cause of many UFO sightings. He discusses a fascinating report on the night of August 13-14, 1960 when two highway patrolmen were chasing a speeding motorcycle when, at about 11:50 pm, they saw what they first supposed to be a brilliantly lighted aircraft falling directly toward them. Jumping out of their car, they watched the object as it apparently reversed its course, shot upward, and began to perform fantastic maneuvers in the eastern sky. The performance continued for more than two hours.

Before it ended, a second UFO had joined in the celestial dance, which was observed by dozens of excited witnesses in the Red Bluff area.

Donald Menzel.

Air Force bases in the area were notified and the Air Force began to investigate the incident. However, as Dr. Menzel tells us, there were no UFOs at all and it was a temperature inversion that caused the star Capella to be brighter than normal plus the planet Mars was rising in the sky about that time as well and the star Aldebaran was just above the horizon and that is what the policemen saw. This is what made them jump out of their patrol car and this is what the thousands of people in Red Bluff, California saw that night.

Essentially Menzel explained all UFO sightings in a similar way, completely ignoring the testimony of trained police officers and pilots and telling them they were actually seeing Mars, Venus, Aldebaran or our Moon zipping and darting around in the sky, sometimes for hours.

Clearly Dr. Menzel knew better but his thankless job was to debunk UFO sightings for the Air Force and try to explain the perceived hysteria of the flying saucer menace. In this role he was an astronomy expert and disinformation propagandist. As a member of the MJ-12 Menzel would have been privy to the most sensitive information about flying saucers and other UFOs, whether they were extraterrestrial visitors or postwar Nazis in their German flying saucers.

According to the biography of Menzel at obscurantist.com, on May 12, 1949, Menzel witnessed a UFO outside Holloman Air Force Base. His original account remained classified until the 1970s. He would later claim that what he saw was a reflection of the moon. Menzel claimed he had another UFO sighting on March 3, 1955 from the North Pole on the daily Air Force Weather "Ptarmigan" flight. He would later claim it was a misidentification

184

of Sirius.

Other members of Majestic-12 consisted of the following individuals:

Lloyd V. Berkner, known for scientific achievements in the fields of physics and electronics, special assistant to the Secretary of State in charge of the Military Assistance Program, executive secretary of what is now known as the Research and Development Board of the National Military Establishment.

Detlev W. Bronk was a physiologist and biophysicist of international repute, chairman of the National Research Council, and a member of the Medical Advisory Board of the Atomic Energy Commission.

Vannevar Bush was a brilliant scientist who joined the Department of Electrical Engineering at Massachusetts Institute of Technology (MIT) in 1919, and founded the company that became the Raytheon Company in 1922. Bush became vice president of MIT and dean of the MIT School of Engineering in 1932, and president of the Carnegie Institution of Washington in 1938. During World War II he headed the U.S. Office of Scientific Research and Development through which almost all wartime military R&D was carried out, including important developments in radar and the initiation and early administration of the Manhattan Project.

Gordon Gray, three times elected to the North Carolina Senate, succeeded Kenneth Royall as secretary of the Army in June 1949.

Dr. Jerome C. Hunsaker, an innovative aeronautical scientist and design engineer, served as chairman of the National Advisory Committee for Aeronautics.

Robert M. Montague was Sandia base commander, Albuquerque, New Mexico, from July 1947 to February 1951.

General Nathan F. Twining, commander of the B-29 superfortresses that dropped the atom bombs on Hiroshima and Nagasaki. In December 1945, he was named commanding general of the Air Material Command headquartered at Wright Field in Dayton, Ohio. In October 1947, he was appointed commander in chief of the Alaskan Command, remaining in that position until May 1950, when he became acting deputy chief of staff for personnel at US Air Force headquarters in Washington, DC.

Sidney W. Souers, a rear admiral, became deputy chief of

General Nathan Twining.

Naval Intelligence before organizing the Central Intelligence Office in January 1946.

Hoyt S. Vandenberg, a much-decorated US Air Force officer, rose to the rank of commanding general of the Ninth US Air Force in France before he was named assistant chief of staff of G-2 (Intelligence) in 1946. In June 1946, he was appointed the director of Central Intelligence.

186

So, we have a mix of admirals, generals, CIA directors, and scientists that are part of this special panel. Most investigators believe that "Document A," which purported to be a letter dated September 24, 1947, from President Harry S. Truman to Secretary of Defense Forrestal, appeared to be genuine; but even though Truman did refer to "Operation Majestic Twelve" in the letter, there was nothing clearly stated that linked the group to UFO investigations.

Admiral Sidney W. Souers.

However, it seems that MJ-12 does have to do with UFOs, but not necessarily the crash at Roswell and the alien bodies supposedly discovered there. It would seem that the UFOs that MJ-12 were looking at were the swarms of Haunebu and Vril saucers being seen around the world plus the Andromeda motherships as well. The CIA warned that flying saucers could be a "psyops" operation being conducted by a foreign power. Such a document would come directly from MJ-12 members Hillenkoetter, Vandenberg, and Souers—all of whom were directors of the CIA at some time in the years immediately after WWII.

Operation Highjump and the German flying saucers encountered there would have alarmed the top Admirals and Generals in the US army and Forrestal was the one who signed the orders for this invasion of Antarctica. It seems that the real purpose of MJ-12 was to be a think tank for what was happening regarding UFOs in the USA and other countries and to control the narrative and somehow deflect attention from the real source of this activity. Menzel's role was clearly to be a leading scientist who could explain to the public that any UFO hysteria was just the misidentification of explainable phenomena. Photos of flying saucers and cylindrical craft were fakes or reflections on windows. This campaign worked relatively well, but popular culture backed by Hollywood did its best to keep flying saucers in the public

187

consciousness.

Did Harry Rositzke know about Project Rainbow and the Philadelphia Experiment? He had probably read Moore and Berlitz's 1979 book plus the other books by Berlitz and Moore including the 1980 book *The Roswell Incident.*[29]

We might conclude that he was interested in the subject but probably had no direct knowledge of the events and only knew what he had read in Moore and Berlitz's book. But Rositzke was a messenger and a high level one at that. We have to conclude that the CIA and Naval Intelligence wanted this information about the existence of MJ-12 to be conveyed to the overall public. This was done via William Moore and Jamie Shandera. Ultimately, we must applaud all of these people, including Rositzke, Admiral Hillenkoetter, William Moore and Jamie Shandera, for getting us some useful information on the government's knowledge of UFOs and secret experiments.

TOP SECRET / MAJIC
EYES ONLY
* TOP SECRET *
****

002

COPY ONE OF ONE.

EYES ONLY

SUBJECT: OPERATION MAJESTIC-12 PRELIMINARY BRIEFING FOR PRESIDENT-ELECT EISENHOWER.

DOCUMENT PREPARED 18 NOVEMBER, 1952.

BRIEFING OFFICER: ADM. ROSCOE H. HILLENKOETTER (MJ-1)

NOTE: This document has been prepared as a preliminary briefing only. It should be regarded as introductory to a full operations briefing intended to follow.

* * * * * *

OPERATION MAJESTIC-12 is a TOP SECRET Research and Development/ Intelligence operation responsible directly and only to the President of the United States. Operations of the project are carried out under control of the Majestic-12 (Majic-12) Group which was established by special classified executive order of President Truman on 24 September, 1947, upon recommendation by Dr. Vannevar Bush and Secretary James Forrestal. (See Attachment "A".) Members of the Majestic-12 Group were designated as follows:

Adm. Roscoe H. Hillenkoetter
Dr. Vannevar Bush
Secy. James V. Forrestal*
Gen. Nathan F. Twining
Gen. Hoyt S. Vandenberg
Dr. Detlev Bronk
Dr. Jerome Hunsaker
Mr. Sidney W. Souers
Mr. Gordon Gray
Dr. Donald Menzel
Gen. Robert M. Montague
Dr. Lloyd V. Berkner

The death of Secretary Forrestal on 22 May, 1949, created a vacancy which remained unfilled until 01 August, 1950, upon which date Gen. Walter B. Smith was designated as permanent replacement.

* TOP SECRET *
TOP SECRET / MAJIC
EYES ONLY

EYES ONLY

T52-EXEMPT (E)

002

Page 2 of the MJ-12 documents.

A-3

TOP SECRET / MAJIC

EYES ONLY

* TOP SECRET *
****

003

EYES ONLY

COPY ONE OF ONE.

On 24 June, 1947, a civilian pilot flying over the Cascade Mountains in the State of Washington observed nine flying disc-shaped aircraft traveling in formation at a high rate of speed. Although this was not the first known sighting of such objects, it was the first to gain widespread attention in the public media. Hundreds of reports of sightings of similar objects followed. Many of these came from highly credible military and civilian sources. These reports resulted in independent efforts by several different elements of the military to ascertain the nature and purpose of these objects in the interests of national defense. A number of witnesses were interviewed and there were several unsuccessful attempts to utilize aircraft in efforts to pursue reported discs in flight. Public reaction bordered on near hysteria at times.

In spite of these efforts, little of substance was learned about the objects until a local rancher reported that one had crashed in a remote region of New Mexico located approximately seventy-five miles northwest of Roswell Army Air Base (now Walker Field).

On 07 July, 1947, a secret operation was begun to assure recovery of the wreckage of this object for scientific study. During the course of this operation, aerial reconnaissance discovered that four small human-like beings had apparently ejected from the craft at some point before it exploded. These had fallen to earth about two miles east of the wreckage site. All four were dead and badly decomposed due to action by predators and exposure to the elements during the approximately one week time period which had elapsed before their discovery. A special scientific team took charge of removing these bodies for study. (See Attachment "C".) The wreckage of the craft was also removed to several different locations. (See Attachment "B".) Civilian and military witnesses in the area were debriefed, and news reporters were given the effective cover story that the object had been a misguided weather research balloon.

****
* TOP SECRET *
****

EYES ONLY TOP SECRET / MAJIC
EYES ONLY

T52-EXEMPT (E)

003

Page 3 of the MJ-12 documents.

TOP SECRET / MAJIC   004
EYES ONLY
****
* TOP SECRET *
****

<u>EYES ONLY</u>   COPY <u>ONE</u> OF <u>ONE</u>.

A covert analytical effort organized by Gen. Twining and
Dr. Bush acting on the direct orders of the President, resulted in a preliminary concensus (19 September, 1947) that
the disc was most likely a short range reconnaissance craft.
This conclusion was based for the most part on the craft's
size and the apparent lack of any identifiable provisioning.
(See Attachment "D".) A similar analysis of the four dead
occupants was arranged by Dr. Bronk. It was the tentative
conclusion of this group (30 November, 1947) that although
these creatures are human-like in appearance, the biological
and evolutionary processes responsible for their development
has apparently been quite different from those observed or
postulated in homo-sapiens. Dr. Bronk's team has suggested
the term "Extra-terrestrial Biological Entities", or "EBEs",
be adopted as the standard term of reference for these
creatures until such time as a more definitive designation
can be agreed upon.

Since it is virtually certain that these craft do not origin-
ate in any country on earth, considerable speculation has
centered around what their point of origin might be and how
they get here. Mars was and remains a possibility, although
some scientists, most notably Dr. Menzel, consider it more
likely that we are dealing with beings from another solar
system entirely.

Numerous examples of what appear to be a form of writing
were found in the wreckage. Efforts to decipher these have
remained largely unsuccessful. (See Attachment "E".)
Equally unsuccessful have been efforts to determine the
method of propulsion or the nature or method of transmission
of the power source involved. Research along these lines
has been complicated by the complete absence of identifiable
wings, propellers, jets, or other conventional methods of
propulsion and guidance, as well as a total lack of metallic
wiring, vacuum tubes, or similar recognizable electronic
components. (See Attachment "F".) It is assumed that the
propulsion unit was completely destroyed by the explosion
which caused the crash.

                ****
                * TOP SECRET *
                ****
<u>EYES ONLY</u> TOP SECRET / MAJIC   T52-EXEMPT (E)
            EYES ONLY   004

Page 4 of the MJ-12 documents.

TOP SECRET / MAJIC
EYES ONLY

0 0 5

****
* TOP SECRET *
****

<u>EYES ONLY</u>   COPY <u>ONE</u> OF <u>ONE</u>.

A need for as much additional information as possible about
these craft, their performance characteristics and their
purpose led to the undertaking known as U.S. Air Force Project
SIGN in December, 1947. In order to preserve security, liason
between SIGN and Majestic-12 was limited to two individuals
within the Intelligence Division of Air Materiel Command whose
role was to pass along certain types of information through
channels. SIGN evolved into Project GRUDGE in December, 1948.
The operation is currently being conducted under the code name
BLUE BOOK, with liason maintained through the Air Force officer
who is head of the project.

On 06 December, 1950, a second object, probably of similar
origin, impacted the earth at high speed in the El Indio -
Guerrero area of the Texas - Mexican boder after following
a long trajectory through the atmosphere. By the time a
search team arrived, what remained of the object had been almost
totally incinerated. Such material as could be recovered was
transported to the A.E.C. facility at Sandia, New Mexico, for
study.

Implications for the National Security are of continuing importance in that the motives and ultimate intentions of these
visitors remain completely unknown. In addition, a significant
upsurge in the surveillance activity of these craft beginning
in May and continuing through the autumn of this year has caused
considerable concern that new developments may be imminent.
It is for these reasons, as well as the obvious international
and technological considerations and the ultimate need to
avoid a public panic at all costs, that the Majestic-12 Group
remains of the unanimous opinion that imposition of the
strictest security precautions should continue without inter-
ruption into the new administration. At the same time, contingency plan MJ-1949-04P/78 (Top Secret - Eyes Only) should
he held in continued readiness should the need to make a
public announcement present itself. (See Attachment "G".)

****
TOP SECRET MAJIC
EYES ONLY

<u>EYES ONLY</u>   T52-EXEMPT (E)

Page 5 of the MJ-12 documents.

TOP SECRET / MAJIC
EYES ONLY

```
***************
* TOP SECRET *
***************
```

00G

EYES ONLY

COPY ONE OF ONE.

ENUMERATION OF ATTACHMENTS:

*ATTACHMENT "A".........Special Classified Executive
Order #092447. (TS/EO)

*ATTACHMENT "B".........Operation Majestic-12 Status
Report #1, Part A. 30 NOV '47.
(TS-MAJIC/EO)

*ATTACHMENT "C".........Operation Majestic-12 Status
Report #1, Part B. 30 NOV '47.
(TS-MAJIC/EO)

*ATTACHMENT "D".........Operation Majestic-12 Preliminary
Analytical Report. 19 SEP '47.
(TS-MAJIC/EO)

*ATTACHMENT "E".........Operation Majestic-12 Blue Team
Report #5. 30 JUN '52.
(TS-MAJIC/EO)

*ATTACHMENT "F".........Operation Majestic-12 Status
Report #2. 31 JAN '48.
(TS-MAJIC/EO)

*ATTACHMENT "G".........Operation Majestic-12 Contingency
Plan MJ-1949-04P/78: 31 JAN '49.
(TS-MAJIC/EO)

*ATTACHMENT "H".........Operation Majestic-12, Maps and
Photographs Folio (Extractions).
(TS-MAJIC/EO)

```
***************
* TOP SECRET *
***************
```

TOP SECRET / MAJIC
EYES ONLY

EYES ONLY

T52-EXEMPT (E)
00G

Page 6 of the MJ-12 documents.

A-8

TOP SECRET

EYES ONLY

THE WHITE HOUSE
WASHINGTON

008

September 24, 1947.

MEMORANDUM FOR THE SECRETARY OF DEFENSE

Dear Secretary Forrestal:

As per our recent conversation on this matter, you are hereby authorized to proceed with all due speed and caution upon your undertaking. Hereafter this matter shall be referred to only as Operation Majestic Twelve.

It continues to be my feeling that any future considerations relative to the ultimate disposition of this matter should rest solely with the Office of the President following appropriate discussions with yourself, Dr. Bush and the Director of Central Intelligence.

[signature: Harry Truman]

TOP SECRET
EYES ONLY

008

Page 7 of the MJ-12 documents.

A-9

Notes by S.T. Friedman: This document was found after a few days of searching in the just declassified boxes of Record Group 341 in Mid 1985 by Jaime Shandera and William Moore. Stanton Friedman had discovered during a visit to the National Archives in March 1985 that the RG was in the process of being classification reviewed. Post cards were received hinting that checking the file would be a good idea. This memo clearly has nothing to do with anything else in Box 189 where it was found. Most likely it was planted there during the classification review which involved many teams of 4 each working for a few weeks in a location where they were able to bring in notes, files, brief cases etc. The item in its original form is a carbon on Dictation Onion Skin by Fox Paper. It is discolored around the edges. My best bet for the actual   July 14, 1954
author is James S. Lay who was Exec. Sec of NSC
and worked very very closely with Cutler and met
"off the Record" with Ike at the WH on July 14, 1954.   TOP SECRET RESTRICTED
The mark through the classification is red 🔾 →   SECURITY INFORMATION

MEMORANDUM FOR GENERAL TWINING

SUBJECT: NSC/MJ-12 Special Studies Project

The President has decided that the MJ-12 SSP briefing should take place during the already scheduled White House meeting of July 16, rather than following it as previously intended. More precise arrangements will be explained to you upon arrival. Please alter your plans accordingly.

Your concurrence in the above change of arrangements is assumed.

Note that the last sentence is almost identical to the wording of another TS Cutler-Twining memo found at the Library of Congress in the Twining papers. 🔾

ROBERT CUTLER
Special Assistant
to the President

Note that there is no signature and no /s/
🔾

NND 857013
5 DTH 1/12/87

COPY
from
E NATIONAL ARCHIVES
ord Group No. RG 341, Records of the Headquarters United
States Air Force

Known as the "Twining Memo" this document seems to confirm the existence of MJ-12 and was discovered in March of 1985.

SECRET
COPY

TSDIN/HMM/ig/6-4100

23 September 1947

TSDIN

SUBJECT: AMC Opinion Concerning "Flying Discs"

TO: Commanding General
         Army Air Forces
         Washington 25, D. C.
         ATTENTION: Brig. General George Schulgen
                    AC/AS-2

1. As requested by AC/AS-2 there is presented below the considered opinion of this Command concerning the so-called "Flying Discs". This opinion is based on interrogation report data furnished by AC/AS-2 and preliminary studies by personnel of T-2 and Aircraft Laboratory, Engineering Division T-3. This opinion was arrived at in a conference between personnel from the Air Institute of Technology, Intelligence T-2, Office, Chief of Engineering Division, and the Aircraft, Power Plant and Propeller Laboratories of Engineering Division T-3.

2. It is the opinion that:

    a. The phenomenon reported is something real and not visionary or fictitious.

    b. There are objects probably approximating the shape of a disc, of such appreciable size as to appear to be as large as man-made aircraft.

    c. There is a possibility that some of the incidents may be caused by natural phenomena, such as meteors.

    d. The reported operating characteristics such as extreme rates of climb, maneuverability (particularly in roll), and action which must be considered evasive when sighted or contacted by friendly aircraft and radar, lend belief to the possibility that some of the objects are controlled either manually, automatically or remotely.

    e. The apparent common description of the objects is as follows:

        (1) Metallic or light reflecting surface.

SECRET
COPY   U-39552

Incl. #1

Air Force memo dated 23 September, 1947 stating that flying saucers are real.

Chapter 9

Secret Technology for the Space Force?

Oh, and there we were in one place
A generation lost in space
With no time left to start again.
—*American Pie*, Don McLean

As we look back at the last days of WWII, Project Rainbow and the genesis of flying saucer technology and high voltage invisibility and teleportation experiments, we can see how it brings us straight through the 1950s with its flying saucer craze and the popular concept of an alien invasion. It was Earth versus the flying saucers in those days while meanwhile the Men in Black were driving their black sedans down country roads ready to warn people about talking too much about the strange things they might have seen.

In the 1960s we had shows like *Star Trek, The Time Tunnel, The Invaders, Hogan's Heroes* (which made the Nazis look like a pretty fun bunch), and a slew of other spy and science fiction shows. Martians couldn't be too far away and time travel was a way to get around in the future. Eventually we had shows and movies like *Star Gate* and the *Star Wars* movies which included jumping through hyperspace to get from one solar system to another. The question is: just how far advanced is the US Navy in its research into these technologies?

There can be no doubt the US Navy, along with the Army and the Air Force—now all combined into the Space Force—would want to achieve the ability to turn ships invisible, to teleport objects around the globe and into space, and to have some sort of anti-gravity flight that would allow submarines to fly and even to travel to other planets. Imagine that: a submarine that begins its journey from an underwater submarine base and then flies through the atmosphere and then voyages into space, visiting the Moon on its way to Mars. That is exactly the vision that the US Navy has had for over 70 years. Have they achieved this? Apparently they have.

No matter the state of the Navy's technology at this point, it can be thought that the genesis was laid with the Project Rainbow Philadelphia Experiments that included degaussing and high voltage experiments on several Navy ships that were wound with copper cables. This is not rocket science. This is Tesla technology, virtually an opposite science.

So, while Nazi flying saucers and cylindrical craft were flying around the world from Antarctica and Tierra del Fuego, the US Navy was conducting their own experiments in anti-gravity and teleportation. They used the invaders from outer space as the cover for their own experimental flights and missions. Meanwhile the postwar SS was flying around the world in a monumental psyop that still has people confused to this day. It wasn't until the document drop in 1984 of the MJ-12 documents that things started to make sense—sort of.

Now in 2025 we have the Space Fleet and its goal of putting astronauts on the Moon and Mars and building bases on both. One has to wonder whether they have not already done this. It may be true that NASA is still waiting to build bases on the Moon but what about the Navy, the Army and the Air Force? Did they ever plan to put bases on the Moon? Yes they did and their plans were top secret for decades. They had plans to put secret military bases on the Moon by 1964 or earlier. They had the rockets and the money to do the job. Did they ever do it? It seems that they probably did. And lessons learned during Project Rainbow and the Philadelphia Experiment helped them to do it.

Solar Warden: The Navy's Secret Space Fleet

A secret space fleet code named Solar Warden was first reported in Britain by the news media when a computer hacker with Aspergers syndrome named Gary McKinnon was "arrested" by the British police in 2002. The basic story is told by British UFO investigator and journalist Darren Perks in an article that appeared in the huffingtonpost.co.uk on November 7, 2012. Said the article:

> ... since 1980, a secret space fleet code named "Solar Warden" has been in operation unknown to the public that allegedly includes key world governments. While conducting an FOI (freedom of information) request with the DOD (department of defense) in 2010, I had a much unexpected response by email from them which read:

>> About an hour ago I spoke to a NASA rep who confirmed this was their program and that it was terminated by then President Obama. He also informed me that it was not a joint program with the DOD. The NASA rep informed me that you should be directed to the Johnson Space Center FOIA Manager.

> The program not only operates classified under the US Government but also under the United Nations authority. So you might be wondering how do I know this information? Well there are a few people and many others that have tried hard to find out the truth, and have succeeded by leaked information or simply asking questions and have government departments slip up and give away information freely, just like what happened when Darren Perks asked the DOD. One notable contributor is Gary McKinnon.

> When Gary McKinnon hacked into U.S. Space Command computers several years ago and learned of the existence of "non-terrestrial officers" and "fleet-to-fleet transfers" and a secret program called "Solar Warden," he was charged by the Bush Justice Department with having committed "the biggest military computer hack of all time", and stood to face prison time of up to 70 years after extradition from UK. But trying earnest McKinnon in open court would involve his testifying to the above-

classified facts, and his attorney would be able to subpoena government officers to testify under oath about the Navy's Space Fleet. To date the extradition of McKinnon to the U.S. has gone nowhere.

McKinnon also found out about the ships or craft within Solar Warden. It is said that there are approximately eight cigar-shaped motherships (each longer than two football fields end-to-end) and 43 small scout ships. The Solar Warden Space Fleet operates under the US Naval Network and Space Operations Command (NNSOC). There are approximately 300 personnel involved at that facility, with the figure rising.

Solar Warden is said to be made up from U.S. aerospace Black Projects contractors, but with some contributions of parts and systems by Canada, United Kingdom, Italy, Austria, Russia, and Australia. It is also said that the program is tested and operated from secret military bases such as Area 51 in Nevada, USA.

So should we just write this off as utter nonsense?

No we shouldn't and as time goes on the truth will slowly come out. Many people around the world are now witnessing craft moving around in the skies and sub space that completely defy gravity. Whether they are part of the Solar Warden secret program, military experimental aircraft or not, thousands of people know what they see.

In my view Solar Warden is very real and a very strong possibility. So no, I don't think we should rule it out as complete nonsense.

Yes, it's a conspiracy because of all the hype and controversy surrounding the facts and information about the program.

Sensitive is an understatement. This program would change the world and our views on space exploration and travel, so no wonder that it would be kept a big 'secret.'

We should all keep it in the back of our minds... for now at least!

According to this article, largely containing information from

Gary McKinnon, we learn that the Solar Warden fleet consists of eight motherships, 43 smaller scout ships and over 300 military personnel, including officers. It is also worth noting that Russia is part of the consortium that also includes Australia, Canada, Italy, Austria and the UK. Notice that neither France nor New Zealand are listed. Nor is Germany. This is a curious collection of countries in some way, considering that New Zealand is part of the Five Eyes intelligence group with the UK, Australia, Canada and the US. Also, it might be noted that the 1989 document dump to Ralf Ettl with the photos and plans of the Haunebu and Vril craft is

Gary McKinnon in 2014.

201

thought to have come from Austria.

An earlier mention of Solar Warden was on the Open Minds Forum in early 2006, though it came anonymously:

> We have a space fleet, which is code named "Solar Warden." There were, as of 2005, eight ships, an equivalent to aircraft carriers and forty-three "protectors," which are space planes. One was lost recently to an accident in Mars' orbit while it was attempting to re-supply the multinational colony within Mars. This base was established in 1964 by American and Soviet teamwork. Not everything is, as it seems.
>
> We have visited all the planets in our solar system, at a distance of course, except Mercury. We have landed on Pluto and a few moons. These ships contain personnel from many countries and have sworn an oath to the World Government. The technology came from back engineering alien-disc wreckage and at times with alien assistance.

The above claims are separate from what Gary McKinnon has said he found on computers at NASA and the Pentagon. Solar Warden does seem to be an international effort of a sort, and they may well have a base on Mars by now. The anonymous author says we have been on the Moon since 1964. One has to wonder if they have already been to Pluto and Neptune; this may just be

The photo of Solar Warden craft in space downloaded by Gary McKinnon.

disinformation. With teleportation technology such as allegedly developed during Project Rainbow such trips to the outer solar system may well be possible.

Gary McKinnon, however, is not disinformation, and Solar Warden was mentioned in British newspapers and other media starting in 2003 because of McKinnon's arrest.

But who is Gary McKinnon? Many Americans have never heard of him, but he is famous across the pond in Great Britain and something of a celebrity.

The Amazing Hacker Named Gary McKinnon

Gary McKinnon has his own Wikipedia page and from it we learn that McKinnon was born on February 10, 1966 and is a:

> ...Scottish systems administrator and hacker who was accused in 2002 of perpetrating the "biggest military computer hack of all time," although McKinnon himself states that he was merely looking for evidence of free energy suppression and a cover-up of UFO activity and other technologies potentially useful to the public.
>
> On 16 October 2012, after a series of legal proceedings in Britain, Home Secretary Theresa May blocked extradition to the United States.
>
> McKinnon was accused of hacking into ninety-seven United States military and NASA computers over a thirteen-month period between February 2001 and March 2002, at the house of his girlfriend's aunt in London, using the name "Solo."

According to Wikipedia, US authorities stated he deleted critical files from operating systems, which shut down the United States Army's Military District of Washington network of two thousand computers for twenty-four hours. McKinnon also posted a notice on the military's website: "Your security is crap." After the September 11 attacks in 2001, McKinnon deleted weapons logs at the Earle Naval Weapons Station, rendering its network of three hundred computers inoperable and paralyzing munitions supply deliveries for the US Navy's Atlantic Fleet. The Pentagon also accused McKinnon of copying data, account files and passwords onto his own computer. US authorities stated that the cost of tracking and correcting the problems he caused was over $700,000.

McKinnon did admit leaving a threat on one computer:

> US foreign policy is akin to Government-sponsored terrorism these days... It was not a mistake that there was a huge security stand down on September 11 last year... I am SOLO. I will continue to disrupt at the highest levels...

US authorities stated that McKinnon was trying to downplay his own actions. A senior military officer at the Pentagon told *The Sunday Telegraph* on July 26, 2009:

> US policy is to fight these attacks as strongly as possible. As a result of Mr. McKinnon's actions, we suffered serious damage. This was not some harmless incident. He did very serious and deliberate damage to military and NASA computers and left silly and anti-America messages. All the evidence was that someone was staging a very serious attack on US computer systems.

McKinnon was first interviewed by the British police on March 19, 2002 and his computer was seized by the authorities. He was interviewed again on August 8 of that year, this time by the UK National Hi-Tech Crime Unit (NHTCU). Then in November 2002, McKinnon was indicted by a federal grand jury in the Eastern District of Virginia. The indictment contained seven counts of computer-related crime, each of which carried a potential ten-year jail sentence. However, McKinnon remained at large for the next three years and frequently talked to the media.

McKinnon told the British media that he obtained unauthorized access to computer systems in the United States including those mentioned in the United States indictment. He told the press that his motivation was drawn from a statement made before the Washington Press Club on May 9, 2001 by "The Disclosure Project" about government suppression of anti-gravity technology.

McKinnon said he hacked into the Pentagon's computers to find evidence of UFOs, antigravity technology, and the suppression of "free energy," something that readers of my books probably know all about. McKinnon has stated that all of these things were behind his actions.

In an interview televised on the BBC's *Click* program on May 5, 2006, McKinnon stated that he was able to get into the military's networks simply by using a Perl script that searched for blank passwords; in other words his report suggests that there were computers on these networks with the default passwords active.

In the 2006 interview with the BBC, he also said of "The Disclosure Project" that "they are some very credible, relied-upon people, all saying yes, there is UFO technology, there's anti-gravity, there's free energy, and it's extraterrestrial in origin and [they've] captured spacecraft and reverse engineered it."

On the BBC program McKinnon said he investigated a NASA photographic expert's claim that at the Johnson Space Center's Building 8, images were regularly cleaned of evidence of UFO craft. He confirmed this, comparing the raw originals with the "processed" images. He claimed to have viewed a detailed image of "something not man-made" and "cigar shaped" floating above the northern hemisphere. Assuming his viewing would be undisrupted owing to the hour, he did not think of capturing the image. He said he was "bedazzled," and therefore did not think of securing it with the screen capture function in the software but then his connection was interrupted. This photo may be the one that can be found on the Internet and is reproduced here. That photo shows five large cylindrical spacecraft and three smaller craft in orbit together above Earth. This is supposedly a blurry photo of a portion of the Solar Warden Space Fleet.

McKinnon continually told the media about the ships or craft within Solar Warden—the eight cigar-shaped Motherships (each longer than two football fields end-to-end) and 43 small "scout ships." He told the media that he stumbled upon a secret list of one hundred off planet US Naval officers and up to ten space warships after breaking into the computer databases. Interestingly a couple of the spaceships were named USSS *Curtis Lemay* and USSS *Roscoe Hillenkoetter*. Both were members of MJ-12.

Gary McKinnon.

McKinnon never really said that he found information about aliens or extraterrestrial contacts. The information that he found was basically about the secret space program. We know that McKinnon is sincere in what he has told the media and he also told the BBC and others that he was afraid that he would end up in Guantanamo Bay if he was to ever stand trial in a US military court.

McKinnon remained at liberty without restriction for three years until June 2005 (when the UK enacted the Extradition Act 2003, which implemented the 2003 extradition treaty with the United States wherein the United States did not need to provide contestable evidence). At that time he became subject to bail conditions, including a requirement to sign in at his local police station every evening and to remain at his home address at night.

If extradited to the US, McKinnon would have faced up to 70 years in jail. When he expressed his fears that he could be sent to Guantanamo Bay on British media he became a celebrity to the British public who felt he was just an innocent young computer genius with Asperger's syndrome who lived with his mom, and not a security threat.

According to Wikipedia, McKinnon's lawyers in the House of Lords on June 16, 2008, told the Law Lords that the US prosecutors had said McKinnon faced a possible 8–10 years in jail per count if he contested the charges (there were seven counts) without any chance of repatriation, but only 37–46 months if he cooperated and went voluntarily to the United States. McKinnon's lawyers contended that in effect this was intimidation to force McKinnon to waive his legal rights. McKinnon also stated that he had been told that he could serve part of his sentence in the UK if he cooperated. McKinnon and his lawyers rejected the offer because the Americans would not guarantee these concessions.

McKinnon's lawyer said that the Law Lords could deny extradition if there was an abuse of process: "If the United States wish to use the processes of English courts to secure the extradition of an alleged offender, then they must play by our rules."

However, the House of Lords rejected this argument, and then McKinnon appealed to the European Court of Human Rights, which briefly imposed a bar on the extradition, but the request for an appeal was rejected. On 23 January 2009, McKinnon won permission from the High Court to apply for a judicial review against his extradition. Upon losing this appeal, McKinnon's legal team applied for a judicial review into the Home Secretary's rejection of medical evidence, which stated that, when he could easily be tried in the UK, it was unnecessary, cruel and inhumane to inflict the further stress of removing him from his homeland, his family and his medical support network. Finally, on October 16, 2012, then-Home Secretary Theresa May announced to the House of Commons that the extradition had been blocked, saying:

> Mr. McKinnon is accused of serious crimes. But there is also no doubt that he is seriously ill [...] He has Asperger's syndrome, and suffers from depressive illness. Mr. McKinnon's extradition would give rise to such a high risk of him ending his life that a decision to extradite would be incompatible with Mr. McKinnon's human rights.

She stated that the Director of Public Prosecutions would determine whether McKinnon should face trial before a British court. On December 14, 2012 it was announced that McKinnon would not be prosecuted in the United Kingdom because of the difficulties involved in bringing a case against him when the evidence was in the United States.

During all these years McKinnon gathered much support from

the public and a number of celebrities. In November of 2008, the rock group Marillion announced that it was ready to participate in a benefit concert in support of McKinnon's struggle to avoid extradition to the United States. Many prominent individuals voiced support, including Sting, Trudie Styler, Julie Christie, David Gilmour, Graham Nash, Peter Gabriel, The Proclaimers, Bob Geldof, Chrissie Hynde, David Cameron, Boris Johnson, Stephen Fry, and Terry Waite. All proposed that, at the very least, he should be tried in the UK.

In August 2009, the Glasgow newspaper *The Herald* reported that the Scottish entrepreneur Luke Heron would pay £100,000 towards McKinnon's legal costs in the event he was extradited to the US. Web and print media across the UK were critical of the extradition, and *The Daily Mail* ran a campaign to prevent it.

Also in August of 2009, Pink Floyd's David Gilmour released an online single, "Chicago—Change the World," on which he sang and played guitar, bass and keyboards, to promote awareness of McKinnon's plight. A re-titled cover of the Graham Nash song "Chicago," it featured Chrissie Hynde and Bob Geldof, plus McKinnon himself. It was produced by long-time Pink Floyd collaborator Chris Thomas and was made with Nash's support.

According to Wikipedia, on July 20, 2010, Tom Bradby, the British broadcaster ITN's political editor, raised the Gary McKinnon issue with President Barack Obama and Prime Minister David Cameron at a joint White House press conference in Washington DC. Obama and Cameron responded that they had discussed McKinnon's situation and were working to find an "appropriate solution."

So, after a decade of international intrigue, the perpetrator of the "biggest military computer hack of all time" could finally rest easy at his mom's house and not have to think about detention in Guantanamo Bay for the rest of his life. But what of Solar Warden? Do we have other evidence that it exists?

Other Evidence for Solar Warden

According to the well-known UFO investigator and writer George Filer, who has written about Solar Warden in a number of his blogs, Navy hospitals have numerous patients claiming they

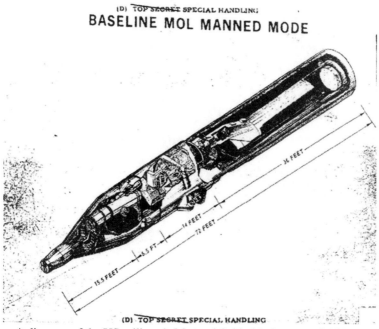

(D) TOP SECRET SPECIAL HANDLING

BASELINE MOL MANNED MODE

(D) TOP SECRET SPECIAL HANDLING

A diagram of the US military's Manned Orbital Laboratory (MOL).

were part of Solar Warden. He believes that it is real. Filer says that because of her support for Gary McKinnon, Theresa May became the new Conservative Party leader and second female prime minister, taking charge of the UK on July 24, 2016. This was because she had become famous in Britain when the Obama Administration was trying to force the UK to extradite McKinnon to face trial in the US. Filer says it was Teresa May who refused to allow McKinnon to be sent to the US and this is what made her popular with the British voting public.

Filer also mentioned that there is an interesting document concerning President Reagan that came to light when the National Archive Records Administration made available 250,000 pages of documents from President Reagan's administration. Filer says that the entry for June 11, 1985 (page 334) reads:

> Lunch with 5 top space scientists.
> "It was fascinating. Space truly is the last frontier and some of the developments there in astronomy etc. are like science fiction, except they are real. I learned that our shuttle capacity is such that we could orbit 300 people."

209

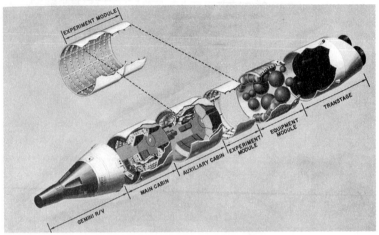

A diagram of the US military's Manned Orbital Laboratory (MOL).

Filer says that this would indicate a space ship at least as large as a 747-8 aircraft that is 250 feet long. It is likely much bigger to accommodate sleeping quarters, kitchen facilities, bathrooms, storage, radar and weapons. The now grounded Space Shuttle held a maximum of eight people and only five were built for space flight.

Filer says that apparently President Reagan revealed the existence of a highly classified space program that could accommodate hundreds of astronauts in orbit. This space program is most probably Solar Warden.

Filer then says that this Navy space fleet could be used to essentially clean up all the space debris that is floating out there in orbit around the Earth, causing potential danger to space stations and satellites, even large debris that NASA or others would normally let fall eventually to Earth. All could be blasted into tiny particles with powerful lasers by the Navy's space force. Says Filer:

> Normally, NASA vehicles would require gradual orbital corrections that would take much time and be insufficient to deal with an immediate threat. According to Ted Twietmeyer, the citation is circumstantial evidence for the existence of antigravity vehicles with advanced particle beam weapons that could remove orbital debris from the

A plastic model of the Manned Orbital Laboratory (MOL).

path of NASA vehicles. These space ships must be armed with particle beam or quantum weaponry, chemical lasers or electrically excited reactor powered lasers and other weaponry. These weapons could also be used to destroy objects in the path of the space station or space shuttle. There were several eyewitness reports of hovering black triangles firing at targets in Iraq during the early days of the war.

So, here we have some corroboration that a space force has existed for many decades. Black triangles with advanced particle beam weapons are apparently part of the Space Force. It is interesting that they need these advanced particle beam weapons to blast orbital debris out of their path as they move through space. This can be space junk as well as micrometeorites as well as other objects. The Federation of American Scientist estimates that there may be as many as 170 million pieces of debris in orbit around the Earth. The European Space Agency estimates that there are 29,000 pieces of debris that are over 10 centimeters in size. This is clearly a serious hazard for any space craft travelling to the Moon or around the Earth. Blasting them with a particle beam seems like a scientific way to clear the way for spacecraft. Would it be possible to just teleport past all these objects?

This brings us to the final question in regard to Project Rainbow and the Philadelphia Experiment: Does the US Navy and the Space

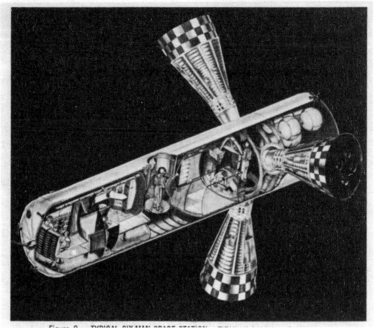

A diagram of the US military's Manned Orbital Laboratory (MOL).

Force have the ability to teleport a ship from one place to another? By extension, can they teleport airship/submarines from one place to another? Also by extension, can they teleport objects to other planets and moons? If they have this ability it is Top Secret and not widely known within the military.

In ending this book I will relate a personal story that occurred in Brisbane, Australia around 1992. I was at a conference and was approached by a middle-aged Australian researcher who wanted to show me some newspaper articles. He also wanted to tell me that he suspected that the US Navy had teleported an aircraft carrier into Brisbane Harbor during an intense storm a couple of years before.

The story he told was of the US putting pressure on Australia to sign a certain UN treaty and this involved President George Bush Sr. flying to Australia. That night there was a terrible storm over Brisbane and the harbor with lots of lightning, thunder, wind and rain. In the morning the storm had cleared up and sitting in the middle of Brisbane Harbor was a US Navy aircraft carrier. He proceeded to show me a newspaper article that detailed how the battleship had suddenly appeared unannounced and was a surprise

212

to the authorities. He also showed me other newspaper articles that said that on that same day the US Navy had told all flights in the mid-Pacific on their way to Australia to avoid a certain area that they normaly flew through and to fly extra miles out of their way—or they would be shot down. The US Navy apparently gave no reason for this diversion of flights.

The Australian researcher believed that the aircraft carrier had been in this region of the central Pacific that was off limits to aircraft and was then teleported to Brisbane Harbor and this also created the amazing lightning storm that was seen on that night. I was fascinated by his theory and research but there was not much else to learn. Was he right about the US Navy teleporting a battleship circa 1989 into Brisbane Harbor? I do not know, but at the time it was very believable and he had some newspaper articles to support his theory, though none spoke about teleportation.

So, in the end it seems probable that Project Rainbow and the Philadelphia Experiment was more than just a degaussing exercise. It also seems that these experiments continued for decades and are ongoing today. It also seems that MJ-12 was a real group that was looking into the flying saucer and UFO phenomenon.

But what of Morris K. Jessup? Why are there apparently two Jessups? Was the real Morris Jessup actual discovered in the car that day? Reportedly his wife refused to look at his body. Was Jessup somehow part of the Rainbow Project in some way? If he was murdered, was it on the orders of Wernher von Braun? Yes, the ghosts of the Philadelphia Experiment continue to haunt us today.

The Pantagraph

Thursday, August 21, 1986

Bloomington-Normal, Illinois

Navy has lots of tricks to make carriers disappear

WASHINGTON, D.C. (AP) — U.S. Navy aircraft carriers, despite their incredible size, are becoming adept at a form of magic.

Utilizing weather, speed, advanced logistical planning and high-tech tomfoolery, several carriers in recent months have managed to vanish from antagonists' eyes into the vastness of the oceans, reappearing only at the moment of attack.

Last April, dogged by airplanes rented by American television networks and by Soviet intelligence networks, the carriers Coral Sea and America dropped from sight off the coast of Sicily. Less than 24 hours later, their planes bombed targets in Libya.

And just over a month ago, a much lengthier case of a "missing" carrier occurred during an exercise named RIMPAC 86. The USS Ranger, although the target of an intense search that included satellite reconnaissance, escaped detection for two weeks while sailing across the Pacific.

The performance was considered all the more remarkable by an Australian admiral who monitored the exercise because the carrier's planes were flying sorties throughout the period, staging mock attacks against surface ships, submarines and land targets.

Rear Adm. I.W. Knox of the Royal Australian Navy disclosed recently the "Or-

ange" forces in RIMPAC could not locate the Ranger "from the time it departed Southern Californian exercise areas until it steamed into Pearl Harbor some 14 days later."

Reports of such exploits delight Navy brass, who must answer critics who think carriers are sitting ducks in an age of nuclear-powered submarines and cruise missiles.

Modern-day carriers have yet to be tested in combat against Soviet weaponry. But they are practicing hard at what the Navy calls "maneuver strategy" — if the enemy can't find you, you have surprise. And with surprise, you can win.

Navy spokesmen decline to discuss the war-fighting tactics, citing military secrecy. But several officers interviewed recently, who asked not to be identified, say the idea of a "stealthy carrier" is not so far-fetched. Consider:

• The Coral Sea and America accomplished their feats through a variety of tricks, but the most important were "masking" and "EMCON." The details of masking are classified, but essentially it involves making another ship — a destroyer, for example — look and sound like a carrier and a carrier look like something else.

Please see CARRIERS, page A5

CARRIERS

From A1

The process normally begins when a carrier is under radar surveillance, but beyond visual sight. The decoy ship maintains the carrier's previous course, while the carrier speeds away.

"We can make the Soviets believe another ship is the carrier," says one official. "The radar image, broadcasting pilot talk and the radio sounds of flight operations, the lighting at night: It looks like a duck and sounds like a duck so it must be a duck. So they follow the duck and make a mistake."

The carrier, meantime, can employ lighting at night that makes it look like a tanker.

• Also employed by the Coral Sea and America, and the key to the Ranger's disappearing act, was EMCON. This is the equivalent of a submarine "rigging for silence" or a convoy traveling under blackout conditions.

EMCON is a Navy acronym for emission control. Emission, in this case, refers to the electronic signals that are radiated by such equipment as radars, sonar and radio. When a carrier goes to EMCON, it literally shuts down much of its electronic gear to avoid detection.

Navy officials say a carrier can operate for long periods in EMCON because "we go mute, but not deaf or blind."

The procedure works by utilizing E2-C Hawkeye radar planes, flying at some distance from the carrier. Everything the Hawkeye sees is relayed electronically to the carrier and its escorts, providing a picture of aerial activity as well as surface forces.

While transmitting, the Hawkeye is far from the carrier, which gets the plane's signals passively without any transmission of its own. The Hawkeye also takes on the role of air-traffic controller for the carrier's planes.

Replenishment oilers, meantime, are told well in advance to make their own way to a specific position in the ocean. Again, radio silence is maintained.

• Aviation tactics. Even if radar can't pick up a carrier sailing beyond the horizon, the ship's location can be betrayed by jet aircraft scrambling into the air. The Navy's answer is called "offset vector."

"To be simplistic, the planes don't climb," says one officer. "They catapult off and literally hit the deck. If planes are suddenly popping up 100 miles from the ship, you have no idea where they came from."

— Speed. Publicly, the Navy says its carriers are capable of speeds "in excess of 30 knots." Privately, officers acknowledge the floating cities can approach 40 knots.

"We can literally outrun the Soviet tattletales (intelligence ships)," says one. "And in (heavy) weather of any kind, there's no contest. The carrier can outrun its own escorts."

— Weather and Satellites. Anyone who's been caught in the rain after the weatherman forecast sunny skies has his own thoughts on meteorology. But there have been solid gains made within that science in recent years.

"Although really heavy weather can hurt flight operations, these guys know how to follow weather patterns and use rain storms and above all, cloud cover," says one official. "The carriers can receive weather data via satellite, passively, without portraying their position."

"And we know the orbital parameters of Soviet reconnaissance satellites as well as our own," adds another. "If there's a recon bird coming by and you can duck into some weather, you duck into the weather. Or if you know there's a blind spot in coverage, you sail there."

"Once you succeed in slipping away," summarizes one officer, "the odds shift in your favor. Most people don't have any conception of how big the oceans are. You can be lonely if you want."

Chapter 10

Bibliography
& Footnotes

1. *Time Travel: Myth or Reality,* Richard Heffern, 1977, Pyramid Publications, New York.

2. *Strange Mysteries of Time and Space,* Harold T. Wilkins, 1958, Citadel Books, New York.

3. *Ancient Astronauts: A Time Reversal?,* Robin Collyns, 1976, Sphere Books, London.

4. *Time Travel: Fact, Fiction & Possibility,* Jenny Randles, 1994, Blandford Books, London.

5. *The Ultimate Time Machine*, Joseph McMoneagle, 1998, Hampton Roads Publishing Company, Charlottesville, Virginia.

6. *Time Machines*, Paul J. Nahin, 1993, Springer Verlag, New York.

7. *The Philadelphia Experiment: Project Invisibility*, Charles Berlitz & William L. Moore, 1979, Grosset & Dunlap, New York.

8. *Without a Trace,* Charles Berlitz, 1977, Doubleday, New York.

9. *The Philadelphia Experiment & Other Conspiracies*, Brad Steiger with Alfred Bielek, 1990, Inner Light Publications, New Brunswick, New Jersey.

10. *The Case for the UFO*, Morris K. Jessup, 1955, Citadel Press, New York.

11. *The UFO Annual,* edited by Morris K. Jessup, 1956, Citadel Press, New York.

12. *UFO and the Bible,* Morris K. Jessup, 1956, Citadel Press, New York.

13. *Behind the Flying Saucers,* Frank Scully, 1950, Fawcett Books, New York.

14. *How to Build a Flying Saucer (And Other Proposals in Speculative Engineering)*, T.B. Pawlicki, 1981, Prentice Hall, Inc., Englewood Cliffs, New Jersey.

15. *How You Can Explore the Higher Dimensions of Space and Time*, T. B. Pawlicki, 1984, Prentice Hall, Inc., Englewood Cliffs, New Jersey.

16. *The Strange Case of Dr. M.K. Jessup*, Gray Barker, 1963, 2014, New Saucerian Press, Point Pleasant, West Virginia.

17. *The Ghost of the Philadelphia Experiment Returns*, Gray Barker, 1981, 2014, New Saucerian Press, Point Pleasant, West Virginia.

18. *The Philadelphia Experiment Chronicles*, Commander X, 1994, Abelhard Publications, Wilmington, Delaware.

19. *NASA, Nazis and JFK: The Torbitt Document and the JFK Assassination*, William Torbitt (pseudonym), 1996, Adventures Unlimited Press, Kempton, IL.

20. *Challenge to Science: The UFO Enigma,* Jacques & Janine Vallee, 1966, Ballantine Books, a Division of Random House, Inc., New York.

21. *Dimensions,* Jacques Vallee, 1988, Ballantine Books, New York.

22. *Messengers of Deception,* Jacques Vallee, 1979, 1980, Bantam Books, Inc., New York.

23. *Anti-Gravity & the World Grid,* David Hatcher Childress, ed., 1987, Adventures Unlimited Press, Kempton, Illinois.

24. *The UFO Encyclopedia,* Jerome Clark, 1996, Visible Ink, Detroit.

25. *Antarctica and the Secret Space Program,* David Hatcher Childress, 2020, Adventures Unlimited Press, Kempton, Illinois.

26. *Haunebu: The Secret Files,* David Hatcher Childress, 2021, Adventures Unlimited Press, Kempton, Illinois.

27. *Andromeda: The Secret Files,* David Hatcher Childress, 2022, Adventures Unlimited Press, Kempton, Illinois.

28. *Vril: Secrets of the Black Sun,* David Hatcher Childress, 2024, Adventures Unlimited Press, Kempton, Illinois.

29. *The Roswell Incident,* Charles Berlitz and William L. Moore, 1980, Grosset & Dunlap, New York.

30. *The Principles of Ultra-Relativity,* Shinichi Seike, 1970, National Space Research Consortium, Uwajima City, Japan.

31. *The Montauk Project,* Preston Nichols and Peter Moon, 1993, Sky Books, Westbury, New York.

32. *Montauk Revisited,* Preston Nichols and Peter Moon, 1994, Sky Books, Westbury, New York.

33. *Pyramids of Montauk,* Preston Nichols and Peter Moon, 1995, Sky Books, Westbury, New York.

34. *Travels Through Time*, Mike Ricksecker, 2023, Connected Universe Press, Cleveland, Ohio.

35. *Time Warps*, John Gribbin, 1979, Dell Publishing Company, New York.

36. *Spaceships in Prehistory*, Peter Kolosimo, 1975, University Books, Seacaucus, New Jersey.

37. *The Philosophy of Time*, Robin Le Poidevin and Murray Macbeath, eds., 1993, MacMillan, New York.

38. *The Assassination of James Forrestal*, David Martin, 2013, Create Space.

39. *The Excalibur Briefing*, Thomas E. Bearden, 1980, Walnut Hill Books, San Francisco.

40. *Time Travelers from Our Future*, Dr. Bruce Goldberg, 1998, Llewllyn Publications, St. Paul, Minnesota.

41. *M.K. Jessup, the Allende Letters, and Gravity*, Riley Crabb, 1962, Borderland Sciences Research Foundation, Vista, California.

42. *Exploring the Physics of the Unknown Universe*, Milo Wolff, 1990, Technotron Press, Manhattan Beach, California.

43. *We Are Not the First*, Andrew Tomas, 1971, Bantam Books, Inc., New York.

44. *Solutions to Tesla's Secrets and the Soviet Tesla Weapons*, Thomas E. Bearden, 1983, Tesla Book Company, Millbrae, California.

45. *A Dual Ether Universe*, Leonid Sokolow, 1977, Exposition Press, Inc., Hicksville, New York.

46. *Stalking the Wild Pendulum,* Itzhak Bentov, 1977, E. P. Dutton, New York.

47. *Alternative (003),* Leslie Watkins, 1977, Avon Books, New York. (Originally a television show in Britain.)

48. *The Energy Grid,* Bruce L. Cathie, 1995, Adventures Unlimited Press, Kempton, Illinois.

49. *The Bridge to Infinity,* Bruce L. Cathie, 1983, Adventures Unlimited Press, Kempton, Illinois.

50. *The Harmonic Conquest of Space,* Bruce L. Cathie, 1998, Adventures Unlimited Press, Kempton, Illinois.

51. *Flying Saucers—Serious Business,* Frank Edwards, 1966, Bantam Books, Inc., New York.

52. *Somebody Else is on the Moon,* George Leonard, 1976, 1977, Pocket Books, a Simon & Schuster Division, New York.

53. *Assassinations,* Nick Redfern, 2020, Visible Ink, Detroit, Michigan.

54. *Who Shot JFK?,* Robin Ramsay, 2002, Pocket Essentials, London.

55. *Beyond Earth: Man's Contact with UFO's,* Ralph Blum with Judy Blum, 1974, Bantam Books, Inc., New York.

56. *Mysticism and the New Physics,* Michael Talbot, 1980, Bantam Books, Inc., New York.

57. *Mysteries of Time and Space,* Brad Steiger, 1974, Dell Publishing Co. New York.

58. *Visitors from Time,* Marc Davenport, 1992, 1994, Greenleaf Publications, Tuscaloosa, Alabama.

59. *New Horizons in Electric, Magnetic and Gravitational Field Theory,* W. J. Hooper, PhD, 1974, Electrodynamic Gravity, Inc., Cuyahoga Falls, Ohio.

60. *Space, Time and Gravitation,* W. Kopczynski and A. Trautman, 1992, MacMillan, New York.

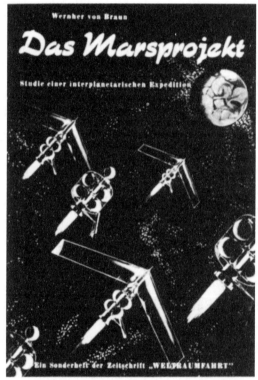

Wernher von Braun's 1949 book *Das Marsprojekt.*

Get these fascinating books from your nearest bookstore or directly from: Adventures Unlimited Press
www.adventuresunlimitedpress.com

COVERT WARS AND BREAKAWAY CIVILIZATIONS
By Joseph P. Farrell
Farrell delves into the creation of breakaway civilizations by the Nazis in South America and other parts of the world. He discusses the advanced technology that they took with them at the end of the war and the psychological war that they waged for decades on America and NATO. He investigates the secret space programs currently sponsored by the breakaway civilizations and the current militaries in control of planet Earth. Plenty of astounding accounts, documents and speculation on the incredible alternative history of hidden conflicts and secret space programs that began when World War II officially "ended."
292 Pages. 6x9 Paperback. Illustrated. $19.95. Code: BCCW

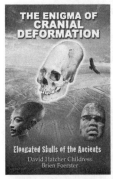

THE ENIGMA OF CRANIAL DEFORMATION
Elongated Skulls of the Ancients
By David Hatcher Childress and Brien Foerster
In a book filled with over a hundred astonishing photos and a color photo section, Childress and Foerster take us to Peru, Bolivia, Egypt, Malta, China, Mexico and other places in search of strange elongated skulls and other cranial deformation. The puzzle of why diverse ancient people—even on remote Pacific Islands—would use head-binding to create elongated heads is mystifying. Where did they even get this idea? Did some people naturally look this way—with long narrow heads? Were they some alien race? Were they an elite race that roamed the entire planet? Why do anthropologists rarely talk about cranial deformation and know so little about it? Color Section.
250 Pages. 6x9 Paperback. Illustrated. $19.95. Code: ECD

ARK OF GOD
The Incredible Power of the Ark of the Covenant
By David Hatcher Childress
Childress takes us on an incredible journey in search of the truth about (and science behind) the fantastic biblical artifact known as the Ark of the Covenant. This object made by Moses at Mount Sinai—part wooden-metal box and part golden statue—had the power to create "lightning" to kill people, and also to fly and lead people through the wilderness. The Ark of the Covenant suddenly disappears from the Bible record and what happened to it is not mentioned. Was it hidden in the underground passages of King Solomon's temple and later discovered by the Knights Templar? Was it taken through Egypt to Ethiopia as many Coptic Christians believe? Childress looks into hidden history, astonishing ancient technology, and a 3,000-year-old mystery that continues to fascinate millions of people today. Color section.
420 Pages. 6x9 Paperback. Illustrated. $22.00 Code: AOG

HESS AND THE PENGUINS
The Holocaust, Antarctica and the Strange Case of Rudolf Hess
By Joseph P. Farrell
Farrell looks at Hess' mission to make peace with Britain and get rid of Hitler—even a plot to fly Hitler to Britain for capture! How much did Göring and Hitler know of Rudolf Hess' subversive plot, and what happened to Hess? Why was a doppleganger put in Spandau Prison and then "suicided"? Did the British use an early form of mind control on Hess' double? John Foster Dulles of the OSS and CIA suspected as much. Farrell also uncovers the strange death of Admiral Richard Byrd's son in 1988, about the same time of the death of Hess.
288 Pages. 6x9 Paperback. Illustrated. $19.95. Code: HAPG

HIDDEN FINANCE, ROGUE NETWORKS & SECRET SORCERY
The Fascist International, 9/11, & Penetrated Operations
By Joseph P. Farrell
Farrell investigates the theory that there were not *two* levels to the 9/11 event, but *three*. He says that the twin towers were downed by the force of an exotic energy weapon, one similar to the Tesla energy weapon suggested by Dr. Judy Wood, and ties together the tangled web of missing money, secret technology and involvement of portions of the Saudi royal family. Farrell unravels the many layers behind the 9-11 attack, layers that include the Deutschebank, the Bush family, the German industrialist Carl Duisberg, Saudi Arabian princes and the energy weapons developed by Tesla before WWII.
296 Pages. 6x9 Paperback. Illustrated. $19.95. Code: HFRN

THRICE GREAT HERMETICA & THE JANUS AGE
By Joseph P. Farrell
What do the Fourth Crusade, the exploration of the New World, secret excavations of the Holy Land, and the pontificate of Innocent the Third all have in common? Answer: Venice and the Templars. What do they have in common with Jesus, Gottfried Leibniz, Sir Isaac Newton, Rene Descartes, and the Earl of Oxford? Answer: Egypt and a body of doctrine known as Hermeticism. The hidden role of Venice and Hermeticism reached far and wide, into the plays of Shakespeare (a.k.a. Edward DeVere, Earl of Oxford), into the quest of the three great mathematicians of the Early Enlightenment for a lost form of analysis, and back into the end of the classical era, to little known Egyptian influences at work during the time of Jesus.
354 Pages. 6x9 Paperback. Illustrated. $19.95. Code: TGHJ

REICH OF THE BLACK SUN
Nazi Secret Weapons & the Cold War Allied Legend
by Joseph P. Farrell
Why were the Allies worried about an atom bomb attack by the Germans in 1944? Why did the Soviets threaten to use poison gas against the Germans? Why did Hitler in 1945 insist that holding Prague could win the war for the Third Reich? Why did US General George Patton's Third Army race for the Skoda works at Pilsen in Czechoslovakia instead of Berlin? Why did the US Army not test the uranium atom bomb it dropped on Hiroshima? Why did the Luftwaffe fly a non-stop round trip mission to within twenty miles of New York City in 1944? Farrel takes the reader on a scientific-historical journey in order to answer these questions. Arguing that Nazi Germany won the race for the atom bomb in late 1944,
352 PAGES. 6x9 PAPERBACK. ILLUSTRATED. $16.95. CODE: ROBS

THE GIZA DEATH STAR REVISITED
An Updated Revision of the Weapon Hypothesis of the Great Pyramid
By Joseph P. Farrell

Join revisionist author Joseph P. Farrell for a summary, revision, and update of his original *Giza Death Star* trilogy in this one-volume compendium of the argument, the physics, and the all-important ancient texts, from the Edfu Temple texts to the Lugal-e and the Enuma Elish that he believes may have made the Great Pyramid a tremendously powerful weapon of mass destruction. Those texts, Farrell argues, provide the clues to the powerful physics of longitudinal waves in the medium that only began to be unlocked centuries later by Sir Isaac Newton and Nikola Tesla's "electro-acoustic" experiments.
360 Pages. 6x9 Paperback. Illustrated. $19.95. Code: GDSR

THE DEMON IN THE EKUR
Angels, Demons, Plasmas, Patristics, and Pyramids
By Joseph P. Farrell

Farrell looks at the Demon in the *Ekur* (the gathering place of the gods in Sumerian tradition) and the Great Pyramid Weapon Hypothesis. He delves deep into the realm of angels, presenting John of Damascus' Angelology and examines the "Immaterial Materiality" of angels; their ability to penetrate ordinary matter and "unlimited" nature; their ability to shapeshift; and the everlasting temporality, or "Sempiternity," of angels. Farrell also explores "Plasma Cosmotheology" and the Plasma Life Hypothesis. He presents intriguing pictures of nuclear detonations and discusses the Hyper-Dimensional Transduction Hypothesis. He includes a discussion of crystals as tuners and transducers, and explores the planetary associations of crystals with angels.
160 Pages. 6x9 Paperback. Illustrated. $16.95. Code: DITE

THE ANTI-GRAVITY FILES
A Compilation of Patents and Reports
Edited by David Hatcher Childress

With plenty of technical drawings and explanations, this book reveals suppressed technology that will change the world in ways we can only dream of. Chapters include: A Brief History of Anti-Gravity Patents; The Motionless Electromagnet Generator Patent; Mercury Anti-Gravity Gyros; The Tesla Pyramid Engine; Anti-Gravity Propulsion Dynamics; The Machines in Flight; More Anti-Gravity Patents; Death Rays Anyone?; The Unified Field Theory of Gravity; and tons more. Heavily illustrated. 4-page color section.
216 pages. 8x10 Paperback. Illustrated. $22.00. Code: AGF

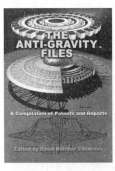

ANCIENT TECHNOLOGY IN PERU & BOLIVIA
By David Hatcher Childress

Childress speculates on the existence of a sunken city in Lake Titicaca and reveals new evidence that the Sumerians may have arrived in South America 4,000 years ago. He demonstrates that the use of "keystone cuts" with metal clamps poured into them to secure megalithic construction was an advanced technology used all over the world, from the Andes to Egypt, Greece and Southeast Asia. He maintains that only power tools could have made the intricate articulation and drill holes found in extremely hard granite and basalt blocks in Bolivia and Peru, and that the megalith builders had to have had advanced methods for moving and stacking gigantic blocks of stone, some weighing over 100 tons.
340 Pages. 6x9 Paperback. Illustrated.. $19.95 Code: ATP

ROSWELL AND THE REICH
The Nazi Connection
By Joseph P. Farrell
Farrell has meticulously reviewed the best-known Roswell research from UFO-ET advocates and skeptics alike, as well as some little-known source material, and comes to a radically different scenario of what happened in Roswell, New Mexico in July 1947, and why the US military has continued to cover it up to this day. Farrell presents a fascinating case sure to disturb both ET believers and disbelievers, namely, that what crashed may have been representative of an independent postwar Nazi power—an extraterritorial Reich monitoring its old enemy, America, and the continuing development of the very technologies confiscated from Germany at the end of the War.
540 pages. 6x9 Paperback. Illustrated. $19.95. Code: RWR

SECRETS OF THE UNIFIED FIELD
The Philadelphia Experiment, the Nazi Bell, and the Discarded Theory
by Joseph P. Farrell
Farrell examines the now discarded Unified Field Theory. American and German wartime scientists and engineers determined that, while the theory was incomplete, it could nevertheless be engineered. Chapters include: The Meanings of "Torsion"; Wringing an Aluminum Can; The Mistake in Unified Field Theories and Their Discarding by Contemporary Physics; Three Routes to the Doomsday Weapon: Quantum Potential, Torsion, and Vortices; Tesla's Meeting with FDR; Arnold Sommerfeld and Electromagnetic Radar Stealth; Electromagnetic Phase Conjugations, Phase Conjugate Mirrors, and Templates; The Unified Field Theory, the Torsion Tensor, and Igor Witkowski's Idea of the Plasma Focus; tons more.
340 pages. 6x9 Paperback. Illustrated. $18.95. Code: SOUF

NAZI INTERNATIONAL
The Nazi's Postwar Plan to Control Finance, Conflict, Physics and Space
by Joseph P. Farrell
Beginning with prewar corporate partnerships in the USA, including some with the Bush family, he moves on to the surrender of Nazi Germany, and evacuation plans of the Germans. He then covers the vast, and still-little-known recreation of Nazi Germany in South America with help of Juan Peron, I.G. Farben and Martin Bormann. Farrell then covers Nazi Germany's penetration of the Muslim world including Wilhelm Voss and Otto Skorzeny in Gamel Abdul Nasser's Egypt before moving on to the development and control of new energy technologies including the Bariloche Fusion Project, Dr. Philo Farnsworth's Plasmator, and the work of Dr. Nikolai Kozyrev. Finally, Farrell discusses the Nazi desire to control space, and examines their connection with NASA, the esoteric meaning of NASA Mission Patches.
412 pages. 6x9 Paperback. Illustrated. $19.95. Code: NZIN

ARKTOS
The Polar Myth in Science, Symbolism & Nazi Survival
by Joscelyn Godwin
Explored are the many tales of an ancient race said to have lived in the Arctic regions, such as Thule and Hyperborea. Progressing onward, he looks at modern polar legends: including the survival of Hitler, German bases in Antarctica, UFOs, the hollow earth, and the hidden kingdoms of Agartha and Shambala. Chapters include: Prologue in Hyperborea; The Golden Age; The Northern Lights; The Arctic Homeland; The Aryan Myth; The Thule Society; The Black Order; The Hidden Lands; Agartha and the Polaires; Shambhala; The Hole at the Pole; Antarctica; more.
220 Pages. 6x9 Paperback. Illustrated. Bib. Index. $16.95. Code: ARK

SAUCERS, SWASTIKAS AND PSYOPS
A History of a Breakaway Civilization
By Joseph P. Farrell
Farrell discusses SS Commando Otto Skorzeny; George Adamski; the alleged Hannebu and Vril craft of the Third Reich; The Strange Case of Dr. Hermann Oberth; Nazis in the US and their connections to "UFO contactees"; The Memes—an idea or behavior spread from person to person within a culture—are Implants. Chapters include: The Nov. 20, 1952 Contact: The Memes are Implants; The Interplanetary Federation of Brotherhood; Adamski's Technological Descriptions and Another ET Message: The Danger of Weaponized Gravity; Adamski's Retro-Looking Saucers, and the Nazi Saucer Myth; Dr. Oberth's 1968 Statements on UFOs and Extraterrestrials; more.
272 Pages. 6x9 Paperback. Illustrated. $19.95. Code: SSPY

LBJ AND THE CONSPIRACY TO KILL KENNEDY
By Joseph P. Farrell
Farrell says that a coalescence of interests in the military industrial complex, the CIA, and Lyndon Baines Johnson's powerful and corrupt political machine in Texas led to the events culminating in the assassination of JFK. Chapters include: Oswald, the FBI, and the CIA: Hoover's Concern of a Second Oswald; Oswald and the Anti-Castro Cubans; The Mafia; Hoover, Johnson, and the Mob; The FBI, the Secret Service, Hoover, and Johnson; The CIA and "Murder Incorporated"; Ruby's Bizarre Behavior; The French Connection and Permindex; Big Oil; The Dead Witnesses: Guy Bannister, Jr., Mary Pinchot Meyer, Rose Cheramie, Dorothy Killgallen, Congressman Hale Boggs; LBJ and the Planning of the Texas Trip; LBJ: A Study in Character, Connections, and Cabals; LBJ and the Aftermath: Accessory After the Fact; The Requirements of Coups D'État; more.
342 Pages. 6x9 Paperback. $19.95 Code: LCKK

THE TESLA PAPERS
Nikola Tesla on Free Energy &
Wireless Transmission of Power
by Nikola Tesla, edited by David Hatcher Childress
David Hatcher Childress takes us into the incredible world of Nikola Tesla and his amazing inventions. Tesla's fantastic vision of the future, including wireless power, anti-gravity, free energy and highly advanced solar power. Also included are some of the papers, patents and material collected on Tesla at the Colorado Springs Tesla Symposiums, including papers on: •The Secret History of Wireless Transmission •Tesla and the Magnifying Transmitter •Design and Construction of a Half-Wave Tesla Coil •Electrostatics: A Key to Free Energy •Progress in Zero-Point Energy Research •Electromagnetic Energy from Antennas to Atoms
325 PAGES. 8x10 PAPERBACK. ILLUSTRATED. $16.95. CODE: TTP

COVERT WARS & THE CLASH OF CIVILIZATIONS
UFOs, Oligarchs and Space Secrecy
By Joseph P. Farrell
Farrell's customary meticulous research and sharp analysis blow the lid off of a worldwide web of nefarious financial and technological control that very few people even suspect exists. He elaborates on the advanced technology that they took with them at the "end" of World War II and shows how the breakaway civilizations have created a huge system of hidden finance with the involvement of various banks and financial institutions around the world. He investigates the current space secrecy that involves UFOs, suppressed technologies and the hidden oligarchs who control planet earth for their own gain and profit.
358 Pages. 6x9 Paperback. Illustrated. $19.95. Code: CWCC

ANTARCTICA AND THE SECRET SPACE PROGRAM
David Hatcher Childress
David Childress, popular author and star of the History Channel's show *Ancient Aliens*, brings us the incredible tale of Nazi submarines and secret weapons in Antarctica and elsewhere. He then examines Operation High-Jump with Admiral Richard Byrd in 1947 and the battle that he apparently had in Antarctica with flying saucers. Through "Operation Paperclip," the Nazis infiltrated aerospace companies, banking, media, and the US government, including NASA and the CIA after WWII. Does the US Navy have a secret space program that includes huge ships and hundreds of astronauts?
392 Pages. 6x9 Paperback. Illustrated. $22.00 Code: ASSP

HAUNEBU: THE SECRET FILES
The Greatest UFO Secret of All Time
By David Hatcher Childress
Childress brings us the incredible tale of the German flying disk known as the Haunebu. Although rumors of German flying disks have been around since the late years of WWII it was not until 1989 when a German researcher named Ralf Ettl living in London received an anonymous packet of photographs and documents concerning the planning and development of at least three types of unusual craft. Chapters include: A Saucer Full of Secrets; WWII as an Oil War; A Saucer Called Vril; Secret Cities of the Black Sun; The Strange World of Miguel Serrano; Set the Controls for the Heart of the Sun; Dark Side of the Moon: more. Includes a 16-page color section. Over 120 photographs and diagrams.
352 Pages. 6x9 Paperback. Illustrated. $22.00 Code: HBU

ANDROMEDA: THE SECRET FILES
The Flying Submarines of the SS
By David Hatcher Childress
Childress brings us the amazing story of the German Andromeda craft, designed and built during WWII. Along with flying discs, the Germans were making long, cylindrical airships that are commonly called motherships—large craft that house several smaller disc craft. It was not until 1989 that a German researcher named Ralf Ettl, living in London, received an anonymous packet of photographs and documents concerning the planning and development of at least three types of unusual craft—including the Andromeda. Chapters include: Gravity's Rainbow; The Motherships; The MJ-12, UFOs and the Korean War; The Strange Case of Reinhold Schmidt; Secret Cities of the Winged Serpent; The Green Fireballs; Submarines That Can Fly; The Breakaway Civilization; more. Includes a 16-page color section.
382 Pages. 6x9 Paperback. Illustrated. $22.00 Code: ASF

VRIL: SECRETS OF THE BLACK SUN
By David Childress
A remnant of the Nazi military—particularly the SS—continued to operate aircraft and submarines around the world in the decades after the end of the war. This volume closes with how the SS operates today in the Ukraine and how the Wagner second in command, Dimitry Utkin, killed in the fiery crash of Yevgeny Prigozhin's private jet between Moscow and St. Petersburg in August of 2023, had SS tattoos on his shoulders and often signed his name with the SS runes. Chapters include: Secrets of the Black Sun; The Extra-Territorial Reich; The Rise of the SS; The SS Never Surrendered; Secret Submarines, Antarctica & Argentina; The Marconi Connection; Yellow Submarine; Ukraine and the Battalion of the Black Sun; more. Includes an 8-page color section.
382 Pages. 6x9 Paperback. Illustrated. $22.00 Code: VSBS

THE GODS IN THE FIELDS

Michael, Mary and Alice-Guardians of Enchanted Britain
By Nigel Graddon

We learn of Britain's special place in the origins of ancient wisdom and of the "Sun-Men" who taught it to a humanity in its infancy. Aspects of these teachings are found all along the St. Michael ley: at Glastonbury, the location of Merlin and Arthur's Avalon; in the design and layout of the extraordinary Somerset Zodiac of which Glastonbury is a major part; in the amazing stone circles and serpentine avenues at Avebury and nearby Silbury Hill: portals to unimaginable worlds of mystery and enchantment; Chapters include: Michael, Mary and Merlin; England's West Country; The Glastonbury Zodiac; Wiltshire; The Gods in the Fields; Michael, Mary and Alice; East of the Line; Table of Michael and Mary Locations; more.
280 Pages. 6x9 Paperback. Illustrated. $19.95. Code: GIF

AXIS OF THE WORLD

The Search for the Oldest American Civilization
by Igor Witkowski

Polish author Witkowski's research reveals remnants of a high civilization that was able to exert its influence on almost the entire planet, and did so with full consciousness. Sites around South America show that this was not just one of the places influenced by this culture, but a place where they built their crowning achievements. Easter Island, in the southeastern Pacific, constitutes one of them. The Rongo-Rongo language that developed there points westward to the Indus Valley. Taken together, the facts presented by Witkowski provide a fresh, new proof that an antediluvian, great civilization flourished several millennia ago.
220 pages. 6x9 Paperback. Illustrated. $18.95. Code: AXOW

LEY LINE & EARTH ENERGIES

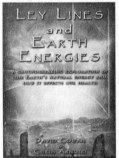

An Extraordinary Journey into the Earth's Natural Energy System
by David Cowan & Chris Arnold

The mysterious standing stones, burial grounds and stone circles that lace Europe, the British Isles and other areas have intrigued scientists, writers, artists and travellers through the centuries. How do ley lines work? How did our ancestors use Earth energy to map their sacred sites and burial grounds? How do ghosts and poltergeists interact with Earth energy? How can Earth spirals and black spots affect our health? This exploration shows how natural forces affect our behavior, how they can be used to enhance our health and well being.
368 pages. 6x9 Paperback. Illustrated. $18.95. Code: LLEE

THE MYSTERY OF U-33

By Nigel Graddon

The incredible story of the mystery U-Boats of WWII! Graddon first chronicles the story of the mysterious U-33 that landed in Scotland in 1940 and involved the top-secret Enigma device. He then looks at U-Boat special missions during and after WWII, including U-Boat trips to Antarctica; U-Boats with the curious cargos of liquid mercury; the journey of the Spear of Destiny via U-Boat; the "Black Subs" and more. Chapters and topics include: U-33: The Official Story; The First Questions; Survivors and Deceased; August 1985—the Story Breaks; The Carradale U-boat; The Greenock Lairs; The Mystery Men; "Brass Bounders at the Admiralty"; Captain's Log; Max Schiller through the Lens; Rudolf Hess; Otto Rahn; U-Boat Special Missions; Neu-Schwabenland; more.
351 Pages. 6x9 Paperback. Illustrated. $19.95. Code: MU33

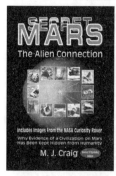

SECRET MARS: The Alien Connection
By M. J. Craig

While scientists spend billions of dollars confirming that microbes live in the Martian soil, people sitting at home on their computers studying the Mars images are making far more astounding discoveries... they have found the possible archaeological remains of an extraterrestrial civilization. Hard to believe? Well, this challenging book invites you to take a look at the astounding pictures yourself and make up your own mind. *Secret Mars* presents over 160 incredible images taken by American and European spacecraft that reveal possible evidence of a civilization that once lived, and may still live, on the planet Mars... powerful evidence that scientists are ignoring! A visual and fascinating book!

352 Pages. 6x9 Paperback. Illustrated. $19.95. Code: SMAR

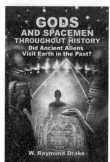

GODS AND SPACEMEN THROUGHOUT HISTORY
Did Ancient Aliens Visit Earth in the Past?
By W. Raymond Drake

From prehistory, flying saucers have been seen in our skies. As mankind sends probes beyond the fringes of our galaxy, we must ask ourselves: "Has all this happened before? Could extraterrestrials have landed on Earth centuries ago?" Drake spent many years digging through huge archives of material, looking for supposed anomalies that could support his scenarios of space aliens impacting human history. Chapters include: Spacemen; The Golden Age; Sons of the Gods; Lemuria; Atlantis; Ancient America; Aztecs and Incas; India; Tibet; China; Japan; Egypt; The Great Pyramid; Babylon; Israel; Greece; Italy; Ancient Rome; Scandinavia; Britain; Saxon Times; Norman Times; The Middle Ages; The Age of Reason; Today; Tomorrow; more.

280 Pages. 6x9 Paperback. Illustrated. $18.95. Code: GSTH

PYTHAGORAS OF SAMOS
First Philosopher and Magician of Numbers
By Nigel Graddon

This comprehensive account comprises both the historical and metaphysical aspects of Pythagoras' philosophy and teachings. In Part 1, the work draws on all known biographical sources as well as key extracts from the esoteric record to paint a fascinating picture of the Master's amazing life and work. Topics covered include the unique circumstances of Pythagoras' birth, his forty-year period of initiations into all the world's ancient mysteries, his remarkable meeting with a physician from the mysterious Etruscan community, Part 2 comprises, for the first time in a publicly available work, a metaphysical interpretation of Pythagoras' Science of Numbers.

294 Pages. 6x9 Paperback. Illustrated. $18.95. Code: PYOS

VIMANA:
Flying Machines of the Ancients
by David Hatcher Childress

According to early Sanskrit texts the ancients had several types of airships called vimanas. Like aircraft of today, vimanas were used to fly through the air from city to city; to conduct aerial surveys of uncharted lands; and as delivery vehicles for awesome weapons. David Hatcher Childress, popular *Lost Cities* author, takes us on an astounding investigation into tales of ancient flying machines. In his new book, packed with photos and diagrams, he consults ancient texts and modern stories and presents astonishing evidence that aircraft, similar to the ones we use today, were used thousands of years ago in India, Sumeria, China and other countries. Includes a 24-page color section.

408 Pages. 6x9 Paperback. Illustrated. $22.95. Code: VMA

HITLER'S SUPPRESSED AND STILL-SECRET WEAPONS, SCIENCE AND TECHNOLOGY
by Henry Stevens

In the closing months of WWII the Allies assembled mind-blowing intelligence reports of supermetals, electric guns, and ray weapons able to stop the engines of Allied aircraft—in addition to feared x-ray and laser weaponry. Chapters include: The Kammler Group; German Flying Disc Update; The Electromagnetic Vampire; Liquid Air; Synthetic Blood; German Free Energy Research; German Atomic Tests; The Fuel-Air Bomb; Supermetals; Red Mercury; Means to Stop Engines; more.

335 Pages. 6x9 Paperback. Illustrated. $19.95. Code: HSSW

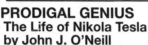
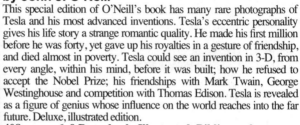

PRODIGAL GENIUS
The Life of Nikola Tesla
by John J. O'Neill

This special edition of O'Neill's book has many rare photographs of Tesla and his most advanced inventions. Tesla's eccentric personality gives his life story a strange romantic quality. He made his first million before he was forty, yet gave up his royalties in a gesture of friendship, and died almost in poverty. Tesla could see an invention in 3-D, from every angle, within his mind, before it was built; how he refused to accept the Nobel Prize; his friendships with Mark Twain, George Westinghouse and competition with Thomas Edison. Tesla is revealed as a figure of genius whose influence on the world reaches into the far future. Deluxe, illustrated edition.

408 pages. 6x9 Paperback. Illustrated. Bibliography. $18.95. Code: PRG

THE ENCYCLOPEDIA OF MOON MYSTERIES
Secrets, Anomalies, Extraterrestrials and More
By Constance Victoria Briggs

Our moon is an enigma. The ancients viewed it as a light to guide them in the darkness, and a god to be worshipped. Did you know that: Aristotle and Plato wrote about a time when there was no Moon? Several of the NASA astronauts reported seeing UFOs while traveling to the Moon?; the Moon might be hollow?; Apollo 10 astronauts heard strange "space music" when traveling on the far side of the Moon?; strange and unexplained lights have been seen on the Moon for centuries?; there are said to be ruins of structures on the Moon?; there is an ancient tale that suggests that the first human was created on the Moon?; Tons more. Tons of illustrations with A to Z sections for easy reference and reading.

152 Pages. 7x10 Paperback. Illustrated. $19.95. Code: EOMM

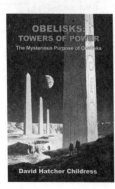

OBELISKS: TOWERS OF POWER
The Mysterious Purpose of Obelisks
By David Hatcher Childress

Some obelisks weigh over 500 tons and are massive blocks of polished granite that would be extremely difficult to quarry and erect even with modern equipment. Why did ancient civilizations in Egypt, Ethiopia and elsewhere undertake the massive enterprise it would have been to erect a single obelisk, much less dozens of them? Were they energy towers that could receive or transmit energy? With discussions on Tesla's wireless power, and the use of obelisks as gigantic acupuncture needles for earth, Chapters include: Megaliths Around the World and their Purpose; The Crystal Towers of Egypt; The Obelisks of Ethiopia; Obelisks in Europe and Asia; Mysterious Obelisks in the Americas; The Terrible Crystal Towers of Atlantis; Tesla's Wireless Power Distribution System; Obelisks on the Moon; more. 8-page color section.

336 Pages. 6x9 Paperback. Illustrated. $22.00 Code: OBK

ORDER FORM

10% Discount When You Order 3 or More Items!

One Adventure Place
P.O. Box 74
Kempton, Illinois 60946
United States of America
Tel.: 815-253-6390 • Fax: 815-253-6300
Email: auphq@frontiernet.net
http://www.adventuresunlimitedpress.com

ORDERING INSTRUCTIONS

✓ Remit by USD$ Check, Money Order or Credit Card

✓ Visa, Master Card, Discover & AmEx Accepted

✓ Paypal Payments Can Be Made To:

    info@wexclub.com

✓ Prices May Change Without Notice

✓ 10% Discount for 3 or More Items

SHIPPING CHARGES

United States

✓ POSTAL BOOK RATE

✓ Postal Book Rate { $5.00 First Item / 50¢ Each Additional Item

✓ Priority Mail { $8.50 First Item / $2.00 Each Additional Item

✓ UPS { $9.00 First Item (Minimum 5 Books) / $1.50 Each Additional Item

NOTE: UPS Delivery Available to Mainland USA Only

Canada

✓ Postal Air Mail { $19.00 First Item / $3.00 Each Additional Item

✓ Personal Checks or Bank Drafts MUST BE US$ and Drawn on a US Bank

✓ Canadian Postal Money Orders OK

✓ Payment MUST BE US$

All Other Countries

✓ Sorry, No Surface Delivery!

✓ Postal Air Mail { $29.00 First Item / $7.00 Each Additional Item

✓ Checks and Money Orders MUST BE US$ and Drawn on a US Bank or branch.

✓ Paypal Payments Can Be Made in US$ To:
    info@wexclub.com

SPECIAL NOTES

✓ RETAILERS: Standard Discounts Available

✓ BACKORDERS: We Backorder all Out-of-Stock Items Unless Otherwise Requested

✓ PRO FORMA INVOICES: Available on Request

✓ DVD Return Policy: Replace defective DVDs only

ORDER ONLINE AT: www.adventuresunlimitedpress.com

10% Discount When You Order 3 or More Items!

Please check: ✓

☐ This is my first order   ☐ I have ordered before

Name

Address

City

State/Province   Postal Code

Country

Phone: Day   Evening

Fax   Email

Item Code	Item Description	Qty	Total

Please check: ✓

Subtotal ▶

Less Discount-10% for 3 or more items ▶

☐ Postal-Surface   Balance ▶

☐ Postal-Air Mail   Illinois Residents 6.25% Sales Tax ▶
(Priority in USA)

     Previous Credit ▶

☐ UPS   Shipping ▶
(Mainland USA only) Total (check/MO in USD$ only) ▶

☐ Visa/MasterCard/Discover/American Express

Card Number:

Expiration Date: Security Code:

✓ SEND A CATALOG TO A FRIEND: